Taming BIG SKY COUNTRY

THE HISTORY OF MONTANA TRANSPORTATION FROM TRAILS TO INTERSTATES

JON AXLINE

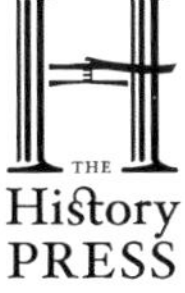
THE History PRESS

Published by The History Press
Charleston, SC 29403
www.historypress.net

Front cover, top: Courtesy of Darrell J. McIntosh. *Bottom*: Courtesy of the Montana Historical Society.

Back cover, top: Courtesy of Darrell J. McIntosh. *Inset*: Courtesy of the Montana Historical Society.

First published 2015

Manufactured in the United States

ISBN 978.1.62619.852.4

Library of Congress Control Number: 2014959300

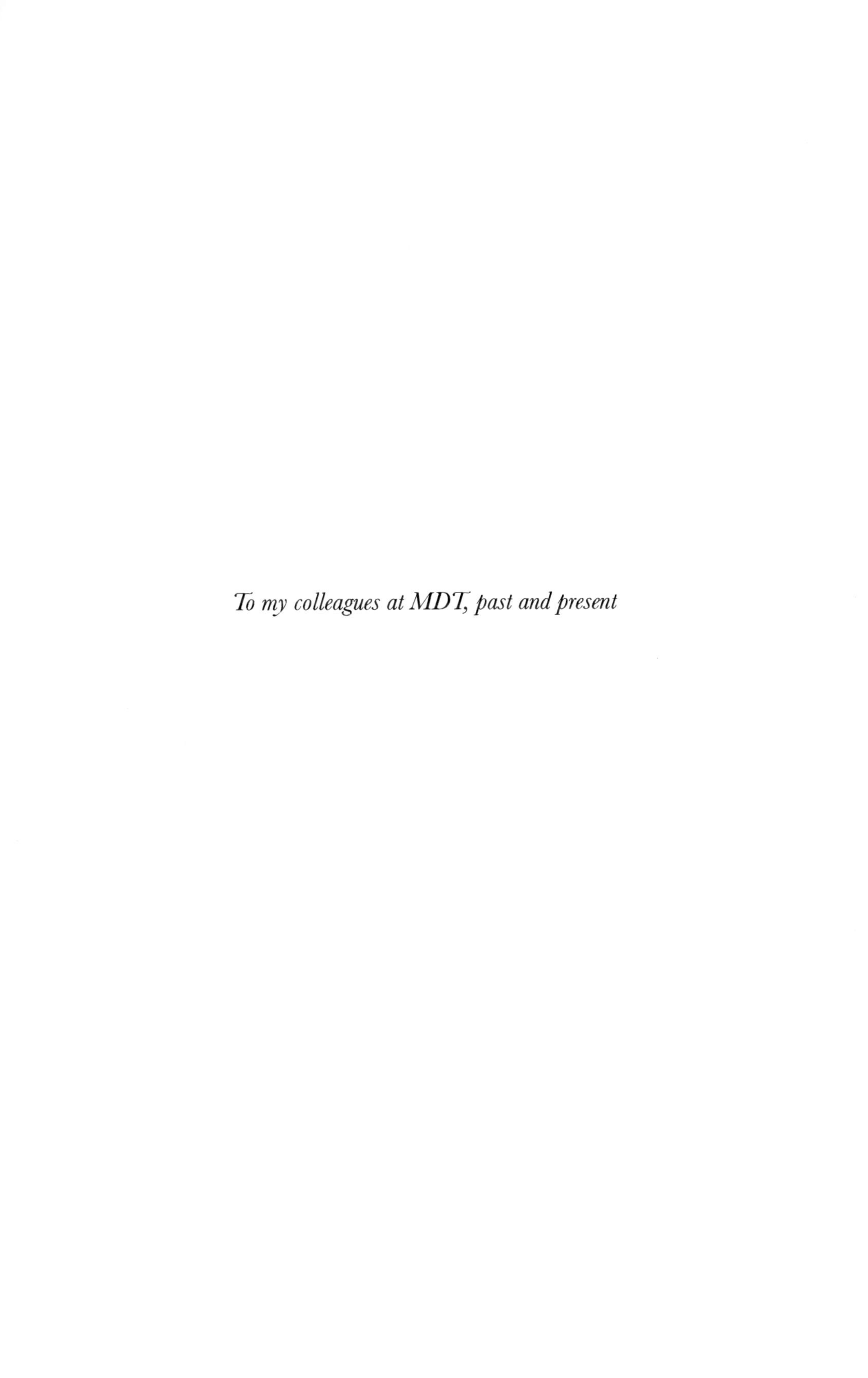

To my colleagues at MDT, past and present

CONTENTS

CONTENTS

ACKNOWLEDGEMENTS

For me, this is the hardest part of the book to write—not because I don't value the people who helped with this project but because it's difficult to think of the right words to describe how much I appreciate them. Writing history, like designing and building a highway, is a team effort; you rely on different sources for information and the help of colleagues and friends to synthesize that information. The Montana Department of Transportation has been my home away from home for three decades. I'm indebted to all my co-workers, especially Theresa Bousliman and Rebecca Goodman, for tolerating a historian in their midst and the tremendous assistance they've provided me.

I'm particularly indebted to Ellen Baumler, Steve Platt, Delia Hagen, Dale Gray, Janene Caywood, Milo MacLeod, the staff at the Montana State Historic Preservation Office and many others for their ongoing efforts to document Montana's history. Finally, my beloved wife, Lisa, has endured the eccentricities of a Montana historian for over twenty-four years. I can never begin to repay all the love and patience she's sent my way over all these years.

Introduction

ROADS TO ROMANCE

Getting Around in Big Sky Country

In 1913, when the state legislature created the Montana State Highway Commission, the state's roads were little more than rutted dirt trails, none of which was constructed to any kind of engineering design standard. Motorists found them difficult, if not impossible, to navigate during much of the year. Unplowed during the winter, the roads were rutted and choked with dust during the summer and quagmires during spring thaws. Travel was an adventure not for the timid.

In 1913, the automobile was in its infancy. The horse-drawn wagons of cowboy artist Charlie Russell's paintings shared the roads with the obnoxious "skunk wagons" of Montana's rapidly increasing urban and rural populations. Although the first highway commission knew that its ultimate goal was the creation of a modern highway system, it was unsure about how to get there with the resources available to it. Montana is a big state with a sparse population; financing and building good roads was a chronic problem.

Montana's road system originated as aboriginal trails used by generations of Native Americans as they followed the bison herds. When Lewis and Clark arrived in what would become Montana in 1805, the explorers noted the presence of Indian "roades" along the river systems they followed to the Pacific coast. The fur trappers and traders later used those trails, and in 1860, Montana's first road engineer, John Mullan, improved a few of them to accommodate wagons. Montana's road system mushroomed in the late nineteenth century with most routes, if not all, following the trails established by the state's first inhabitants many years before.[1]

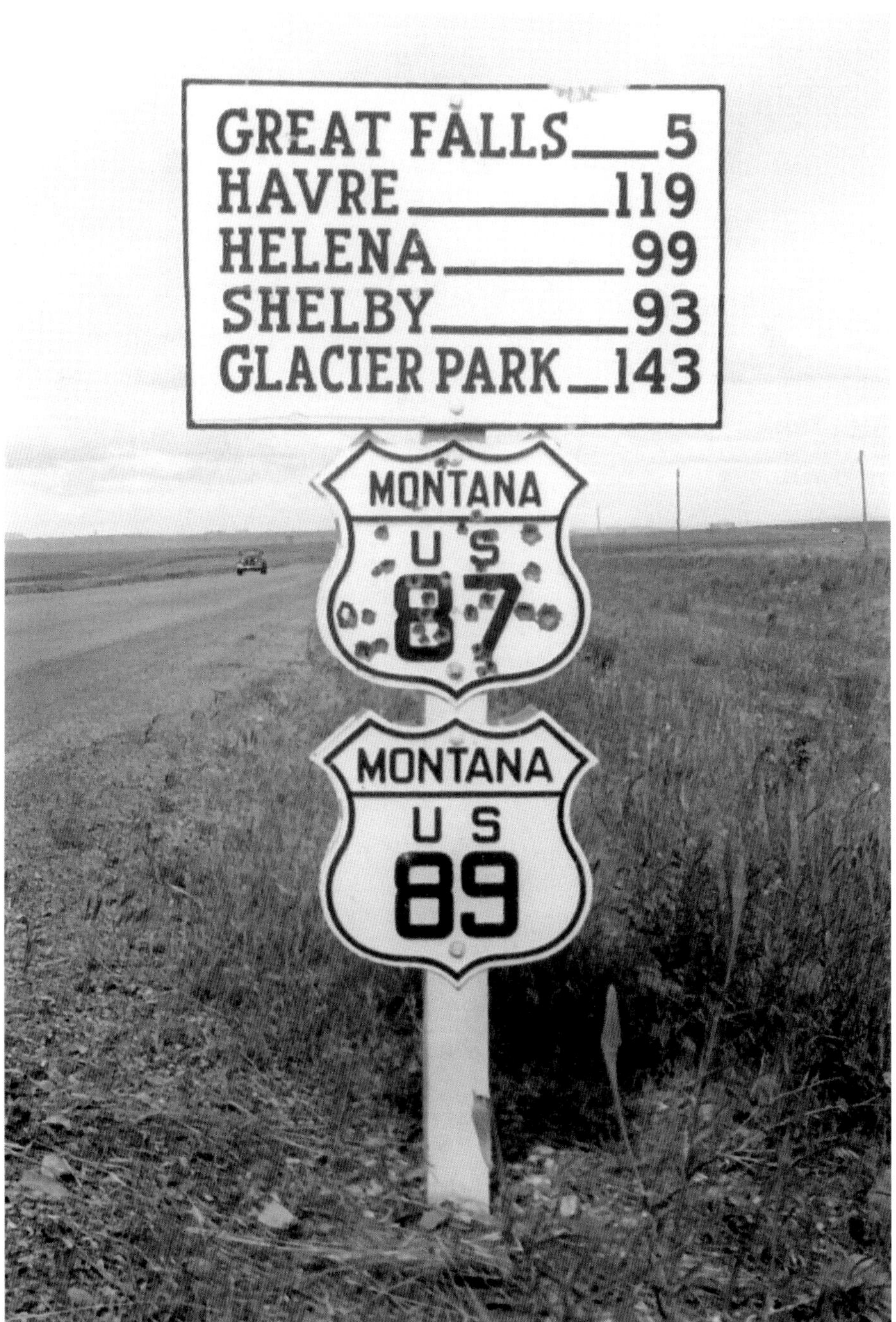

A bullet hole–riddled highway mileage sign. *MDT.*

After decades of local road-building efforts that suited the wagon traffic of the time, the arrival of automobiles in the early twentieth century forced the counties and the state to consider a modern network of roads to accommodate Montana's rapidly growing population and economy. When the Thirteenth Montana Legislature created the state highway commission in March 1913, it could not have envisioned where that legislation would take the state. In less than fifty years, Montana's roads went from primitive two-track dirt paths to paved four-lane superhighways.

With increasing federal involvement in road building in the twentieth century came the added responsibility of the state to maintain the system. Other issues involved right of way acquisition, the environment, politics and finances. Montana chose to build its highways on a pay-as-you-go system, unlike other states that opted for issuing bonds to construct roads. The Treasure State, therefore, didn't go into debt for its highways and bridges like other states, but Montana's sparse population and relatively small gasoline tax revenues made it difficult to match federal funds when they became available to the state. The state developed creative funding methods to accomplish the goal of a modern highway system, paid for mostly with federal dollars.

Today, building highways is a group effort. It requires planners, surveyors, engineers, designers, right of way agents, wildlife biologists, archaeologists, historians, hazardous material specialists, accountants, administrators and contractors. Without them, the projects could never be built, and the Montana Department of Transportation would fail in its mandate to provide "a transportation system and services that emphasize quality, safety, cost-effectiveness, economic vitality, and sensitivity to the environment." In the early days of the Montana Highway Department, however, it was primarily an engineers' domain. Roads were built on alignments that best suited the design standards of the day with no real thought about the consequences. Admittedly, some of the decisions were not good ones, but the engineers succeeded in doing what they set out to do: provide Montanans with modern highways.

This is the story of Montana's highways from their origins as aboriginal trails to the modern multi-lane paved facilities of today. There is no attempt to hide the warts that attended that process because they are important in understanding how the tale unfolded. Transportation was a priority of the first territorial legislators in 1864 and continues to be a priority for the state's legislators in the twenty-first century. How we got from there to here is important to our understanding of Montana's history and the direction we will go in the future.

MONTANA'S ROADS AND TRAILS, 1867

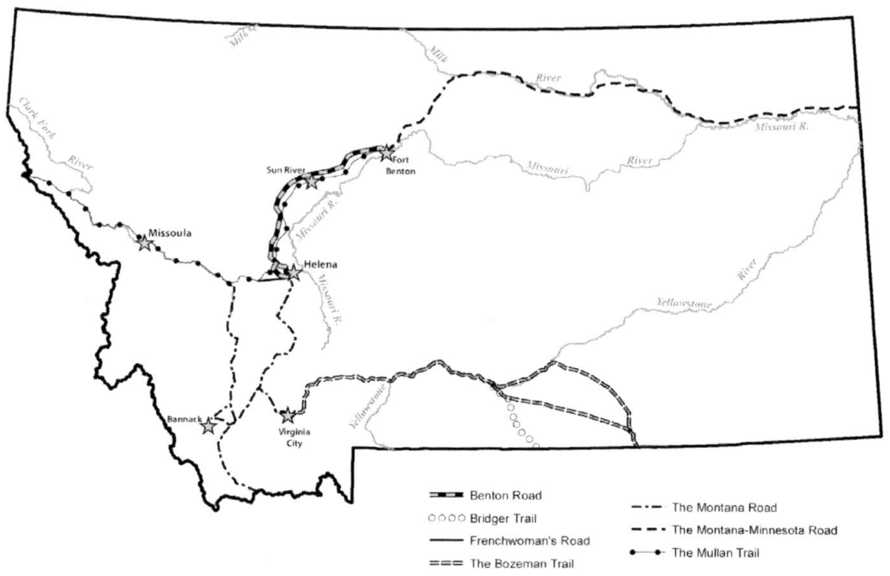

Montana Territory's Roads and Trails, 1867.

Part I

ENTERPRISES OF INCALCULABLE IMPORTANCE

Montana's Early Roads

For a short time in the late nineteenth century, Montana's roads resonated with the sounds of snapping bullwhips, creaking wagon wheels and the profanity-laced encouragements of bullwhackers, muleskinners and stagecoach drivers of oxen, mules and horses. Roads followed hard on the heels of the miners, merchants, saloonkeepers and hurdy-gurdy girls who stampeded to the Montana gold mines beginning in 1862. Just as important as extracting the precious metal from the stream and hillsides was the need for bringing supplies to the mining camps and exporting the gold back to the states. It wasn't long before the primitive aboriginal trails described by Lewis and Clark in 1805 were clogged with wagons, stagecoaches and whatever conveyance best suited the pilgrims attempting to reach the mines. Today's paved highways follow the routes established by our frontier forebears in the 1860s.

To paraphrase Topsy in Harriett Beecher Stowe's *Uncle Tom's Cabin*, Montana's first roads weren't built, they just kind of growed. Except for the Mullan Road, trails developed over time as a result of use. None was surveyed or built to any kind of engineering standard. They mostly occupied the line of least resistance and took meandering routes to avoid substantial geographic obstacles rather than conquer them. Tollkeepers undertook most of the construction and maintenance. They simply rolled troublesome rocks off to the side of the road and cut down trees, leaving just enough clearance for the wagons to pass unimpeded over the stumps. Some builders corduroyed low spots and widened too-narrow places by cutting into the hillsides. They

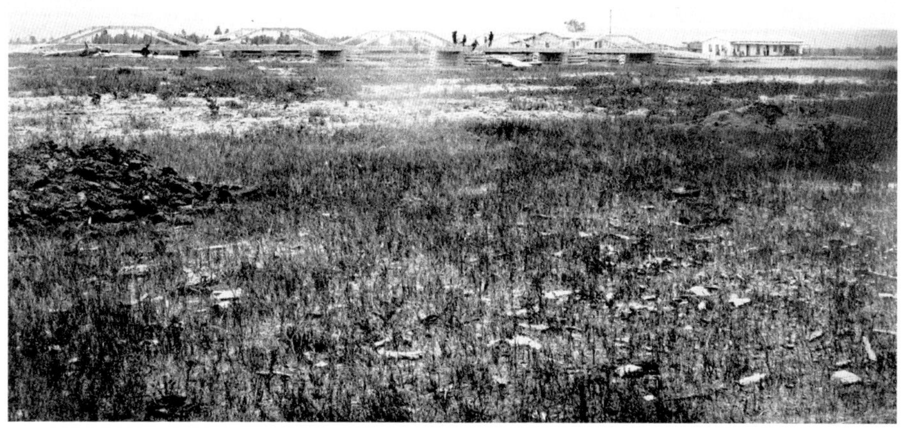

The Sun River toll bridge on the Benton Road, established in 1867 by John Largent and Johnny Healey. *MHS Photograph Archives, Helena, 951-193.*

did much of the work with dollar signs in their eyes. Montana's early roads were literally a pay-as-you-go system—toll collectors built the roads and charged travelers for their use. Roads and bridges were lucrative concerns for tollkeepers and shortcuts common as the frontier impresarios sought ways to undercut their competitors. By the early 1870s, a tangled system of roads and bridges crisscrossed Montana Territory. A few roads predated the gold rush, but most came later. All were, essentially, branches of two roads: the Mullan Road and the Montana Road. Both originated as ancient aboriginal trails.

The Mullan Road

In 1853, Congress allocated money for the surveying of potential routes for a transcontinental railroad in the American West. Secretary of War Jefferson Davis assigned Washington territorial governor Isaac Stevens the task of finding a route for the railroad in the Pacific Northwest and northern Rocky Mountains. For five years, he and his subordinates surveyed the region. They mapped mountain passes, charted the best valleys for a railroad and signed treaties with the Salish, Pend d'Oreille, Kootenai, Blackfeet and Gros Ventre people to allow roads across their territories. In 1854, Lieutenant John Mullan joined Stevens's mission. A recent graduate of West Point, Mullan had a flair for surveying and exploration. It soon became obvious to Stevens that the ambitious young officer had more than just a railroad on his mind. Although Mullan trusted in the railroads as avatars of civilization, he was just as sure of the advantages of an engineered wagon road to serve Manifest Destiny.

During the winter of 1854, Stevens instructed Mullan to seek "routes practicable for a…wagon road" across the northern Rocky Mountains between Walla Walla, Washington, and the head of navigation on the upper Missouri River, Fort Benton. Four years later, in 1858, Captain Andrew Humphreys of the U.S. Army's Office of Exploration and Surveys ordered Mullan to construct the road. The Cayuse War delayed the start of work on the road for a year. The interruption, however, gave now Congressman Stevens and Mullan the opportunity to secure $100,000 from Congress to finance construction of the proposed road.[2]

Mullan and a 230-man crew began work on the military road in July 1859 at Walla Walla. He utilized existing aboriginal trails as much as possible, but rugged terrain in northern Idaho and western Montana made construction of new road segments necessary. The work crew crossed over the Bitterroot Mountains into Montana in early December 1859. Harsh weather conditions soon forced road construction to cease for the season, and Mullan

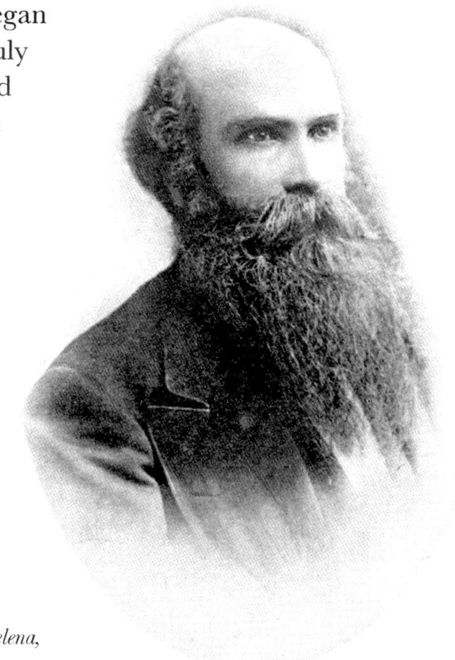

John Mullan. *MHS Photograph Archives, Helena, 944-034.*

established an encampment, Cantonment Jordan, in the St. Regis River Valley to wait out the winter and plan for the next year's construction.

Work resumed on the road in mid-March 1860. Within a month, the company reached a mountain spur, just west of today's Alberton, that extended down to the edge of the Clark Fork, making a road along the riverbank impossible. Mullan later stated that "in order to obtain the practicable elevation on account of the abrupt rocky face of the spurs, I carried the line up a ravine, until gaining 1,000 feet; I wound around the mountain sides, making the re-entering angles by gentle curves, until the entire six miles was completed." Called the "Big Side Cut" and "Point of Rocks" segments in Mullan's report to Congress, he detailed 150 men to work on the six-mile-long detour across the mountainsides. Construction on the segments began on May 1, 1860, and continued for the next six weeks. Because of the rocks along the planned route, the work crew was forced to blast passages through some of the outcrops to maintain Mullan's "gentle curves." Randall Hewitt reported in 1862, "Not an inch more rock was removed than apparently necessary," and the cuts were so narrow that one could not walk next to a wagon while passing through them.[3]

After reaching the Missoula Valley, construction of the road progressed rapidly as Mullan utilized well-worn Indian trails. On July 29, 1860, the expedition arrived in Fort Benton. While the project had gone over budget by then, Mullan completed it by the government-imposed deadline. Upon Mullan's arrival in Fort Benton, he was met by a detachment of soldiers under the command of Major George Blake, who planned to travel back to Walla Walla over the newly completed road. With Mullan improving and repairing the road in advance of the expedition, it traversed the road to Walla Walla in fifty-seven days. The Blake Expedition was the only instance where the military used the road for its intended purpose: to enable the movement of troops from the head of navigation on the Missouri River to the Pacific Northwest. Mullan continued to make improvements to the road until 1862, when the federal government withdrew its financial support because of the pressing demands of the Civil War.

The Mullan Road was the first engineered road constructed in what would become Montana. The 624-mile route traversed some of the most rugged territory in the West. Mullan was justifiably proud of the road and promoted it at every opportunity, including publishing a miners and travelers' guide in 1865. Unfortunately, the road was not entirely a success. The road west of the Missoula Valley passed through particularly inhospitable terrain. Within a year of its completion, fallen timber, forest fires, rock falls and steep grades

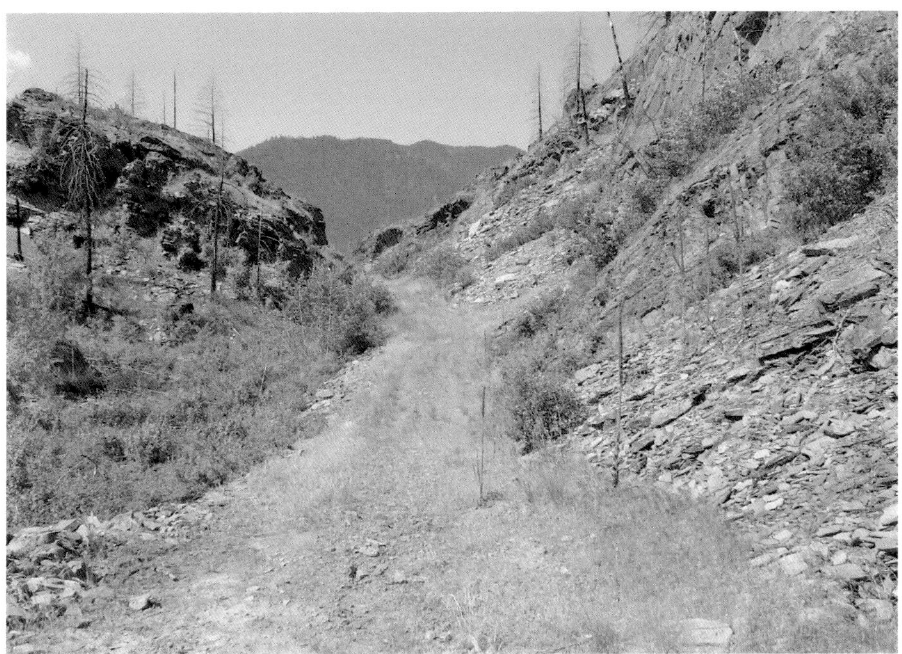

The Point of Rocks segment of the Mullan Road displays the difficulties Mullan experienced while building the road. *MDT.*

prevented its use as a supply route between western Montana and the Pacific Northwest. With the discovery of gold in southwestern Montana in 1862, however, the road east of Missoula became a heavily traveled freight and immigrant road. The Mullan Road between Fort Benton and Helena was known by a different name: the Benton Road.

The Montana Road

The real commerce between Montana and the states in the 1860s passed over a trail that was much more mundane than the Mullan Road. Located on a travel route used by Native Americans for thousands of years, the Montana Road was the territory's most important and well-used lifeline to Utah and the East. The route, which today parallels Interstate 15 between Tremonton, Utah, and Monida Pass, had been used by Euro-Americans since at least 1841, when the Bidwell Expedition explored the Snake River Plain and crossed Monida Pass into what would become southwest Montana. Eight years later, in 1849, Captain J. Howard Stansbury returned from a military reconnaissance between Salt Lake City and Fort Hall and declared it the "best natural road I ever saw." In the 1850s, Richard and Johnny Grant obtained worn-out livestock from migrants on the Oregon Trail at Fort Hall in Idaho and drove the animals north over the road into the Beaverhead and Deer Lodge Valleys in Montana. After fattening up the cattle for a year, the Grants drove them south and traded the healthy animals for worn-out stock. By the early 1860s, travelers had no problem following the road.[4]

Miners and businessmen stampeded to rich gold placers on Grasshopper Creek in 1862 and Alder Gulch in 1863. The road was the link between the Overland Trail in Utah and the new mining camps. Salt Lake City was the original southern terminus of the road, but with the completion of the Union Pacific's transcontinental railroad in 1869, the terminus moved north to the rail station at Corinne. From there, countless tons of freight and thousands of migrants made their way north into Montana. The route to the gold fields crossed a series of high plateaus connected by narrow canyons and low divides in eastern Idaho. The relative ease of the road made it ideal for freight trains and stagecoaches. The Earl of Dunraven described the road to Montana in 1874:

> *Nothing is more extraordinary and wearisome than the levelness of the road. From Corinne to Virginia City you drive along a series of apparently perfectly flat plains, connected with each other by short canyons and valleys. Occasionally the road ascends, but by a very easy gradient. There are no precipices, no torrents, no avalanches, no glaciers; nothing grand, terrible or dangerous. The idea that you are surmounting a portion of a great and important watershed, that you are crossing the backbone of the continent, and scaling a vast mountain range, appears preposterous.*[5]

Because of the easy terrain over which the road passed, freight wagon trains could make four round trips per season, while stagecoaches could make the trip from Corinne to Virginia City in seventy-two hours.

Once the road crossed into Montana over Monida Pass, it branched to different destinations. The main route continued north along the Red Rock and Big Hole Rivers, crossing the Continental Divide north of Melrose, and then continued on to the settlement of Silver Bow with turnoffs east to Butte and west to the Deer Lodge Valley. South of present-day Dillon, the road moved up the Beaverhead River to Beaverhead Rock, where it forked northeast to Helena and the Benton Road; the other branch terminated at Virginia City.

The Bannack–Virginia City segment of the road is the site of many of the events associated with Montana's legendary vigilantes. Most of the freight destined for Montana's mining camps came over the Montana Road.

Long trains of freight wagons were a common sight on the Montana Road between 1864 and 1883. The enormous freight wagons creaked and groaned down the roads at the walking pace of the oxen teams pulling them. Bullwhackers prodded the animals along with bullwhips and colorful language. Some old-timers later remembered hearing the bullwhackers long before sighting the wagons. Each wagon set consisted of three wagons coupled together, with the largest wagon, sometimes with wheels seven feet in diameter, bringing up the rear. Trains averaged around twenty-five sets of wagons, each wagon overseen by a driver, who walked alongside it. Six or seven yoke of oxen towed each wagon set. Each team of the slow but reliable oxen could haul as much as twenty thousand pounds of supplies.

If hauling perishable goods, the freighters relied on the faster mule teams. In rugged areas, like the Mullan Road west of Missoula, pack mules and, occasionally, camels carried supplies to the mining camps

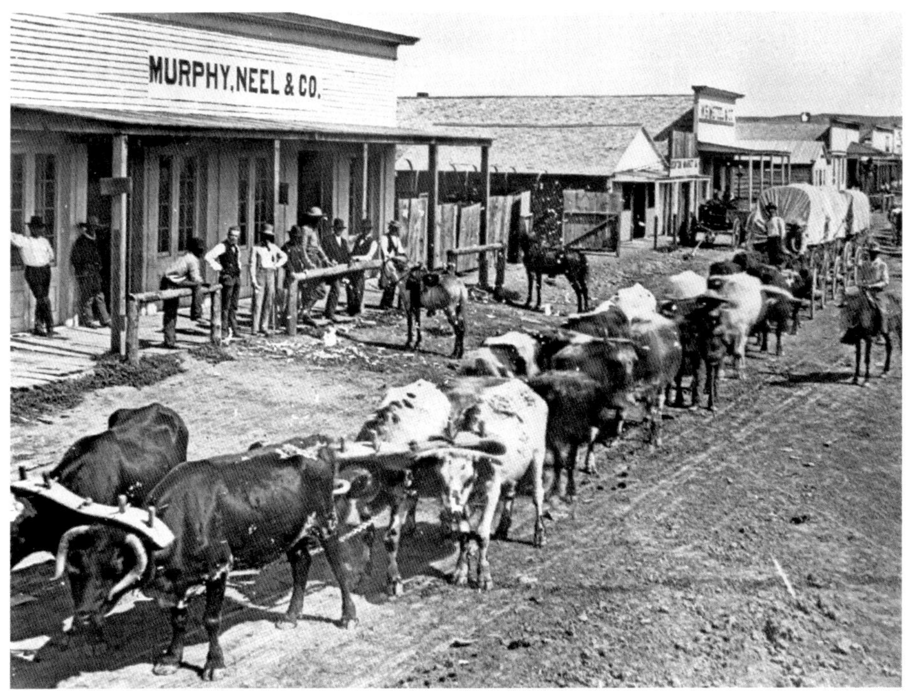

Ox trains, like this one at Fort Benton, were a common sight on Montana's roads. *MHS Photograph Archives, Helena, 947-075.*

A Monument Above the Water: Parson's Bridge

Bridges were critical to Montana Territory's early transportation system. The first territorial legislature chartered seven bridge companies, permitting them to charge travelers for their use. Montana history's most well-known toll bridge was Parson's Bridge. Born in New York about 1825, Nelson Parson arrived in Alder Gulch in 1864. Like many immigrants to the gold fields, he worked in the placer mines on the gulch for a time before seeking other, more profitable prospects. Parson recognized the opportunities afforded by transportation in the territory. In 1865, the first territorial legislature licensed the Jefferson Bridge Company to construct a span across the Jefferson River on the road between Alder Gulch and Helena. Shortly after receiving the territorial license, Parson established a ferry at this crossing of the Jefferson River. The *Montana Post* described Parson's ferry as "one of the finest and best arranged means of transportation in this territory." The following year, in 1866, Parson and several hired hands probably built the bridge shown in the photograph.[6]

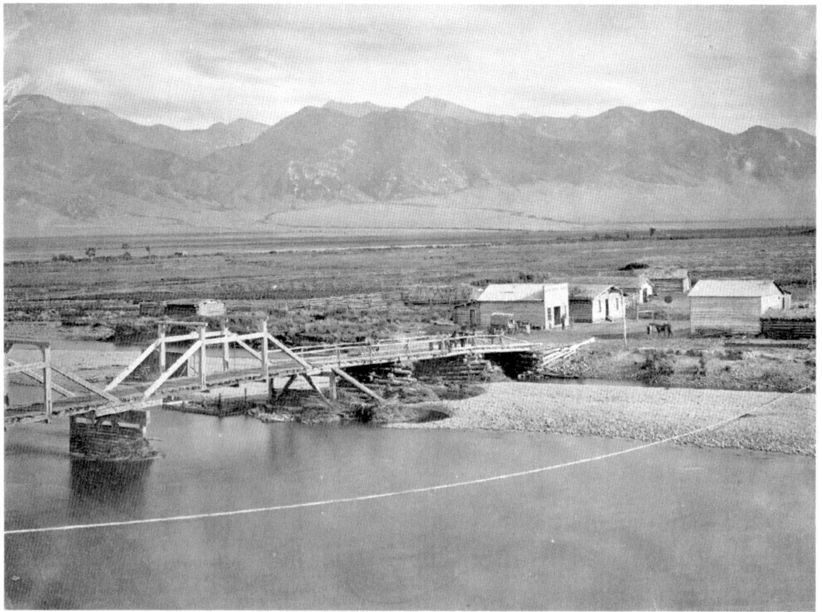

Parson's Bridge crossed the Jefferson River on the road between Virginia City and Helena. *MHS Photograph Archives, Helena, 957-841.*

Once he completed the bridge, Parson charged the same rates authorized by the territorial legislature: five dollars for each wagon and yoke of oxen and fifty cents for each horseman; people on foot could use the bridge for free. By 1868, a small settlement called Jefferson Bridge, comprising half a dozen buildings, had grown up adjacent to the bridge on the south bank of the river. It consisted of a general store, a saloon and several cabins. The settlement obtained a post office in 1866 and, after a few deaths occurred in the area, a cemetery. Parson's wife collected tolls while he farmed and ranched in the area. Parson's toll bridge did good business for a few years, but Madison County took control of the bridge in 1870 and stopped charging tolls for its use. The county replaced the bridge in 1873 after spring floods washed away the original structure. Today, a modern pre-stressed concrete bridge crosses the Jefferson River at Parson's crossing. Amazingly, a small settlement still exists on the south bank of the Jefferson River near the existing bridge.

and settlements. By the mid-1860s, the Montana and Benton Roads were clogged with freight traffic. During the 1866 travel season, an estimated 2,500 wagons pulled by twenty thousand oxen and driven by three thousand bullwhackers traversed the Benton Road over a five-month period beginning in the spring of that year.

The Bozeman Trail

Some Montana roads are deeply entrenched in the folklore of the American West. In 1863, John Bozeman and mountain man John Jacobs blazed a shortcut to the Montana gold fields from the Overland Trail in central Wyoming. At first, it seemed like a good idea to the trailblazers and their partners in the Gallatin Valley of southwestern Montana. It would allow miners, merchants and settlers to reach the gold camps more quickly than the Montana Road—and it passed through the Gallatin Valley, providing an opportunity for businessmen there as well. One thing the partners hadn't counted on was the violent opposition to the route by the Lakota and Northern Cheyenne Indians, whose land it crossed. Roads and buffalo didn't mesh very well, and the Indians knew that. Within

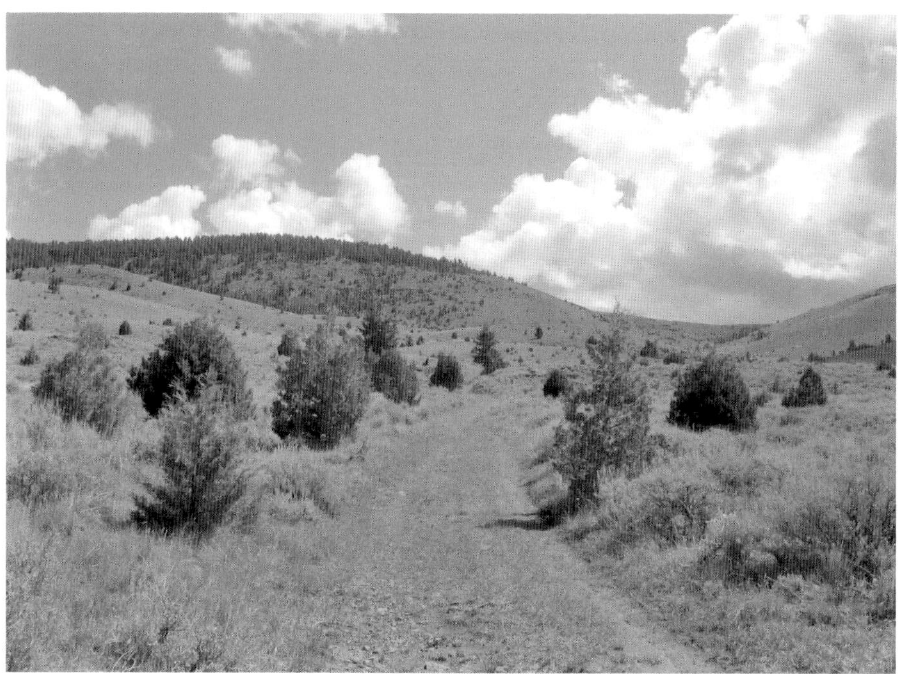

The Bozeman Trail brought hundreds of fortune hunters to Montana Territory between 1864 and 1868. This segment is located in Virginia City. *MDT.*

months of the establishment of Bozeman's Trail, the Indians began to turn back or attack the wagon trains on the trail. Outcry from settlers in Montana Territory compelled the military to establish three forts along the trail in 1866. The tribes promptly placed the posts under siege, effectively closing the trail to all traffic by 1867. In 1866 and 1867, the Fetterman Fight, Wagon Box Fight and Hayfield Fight demonstrated the tribes' willingness to defend their territory and keep out the invaders.

Famed mountain man Jim Bridger proposed an alternate route west of the Big Horn Mountains in 1864. What the Bridger Cutoff lacked in water and grass resources it made up for in the safety of the sojourners willing to travel that way. The cutoff never really caught on, and Bridger had largely abandoned it by 1867. In 1868, the U.S. Army negotiated a treaty with the Lakota and Northern Cheyenne Indians at Fort Laramie. In it, the federal government agreed to abandon its military posts and shut down the trail. This was the only instance in which the Indians forced concessions from the U.S. government. The Bozeman Trail between the Gallatin Valley and Virginia City prospered as an important trade route.

The Northern Overland Route

Gold drew thousands of people to Montana beginning in 1862. The remoteness of the territory, however, limited the paths to the mining camps. During the early years of the Montana gold rush, time was a critical factor, and faraway Montana wasn't easy to get to. Consequently, travelers were always on the hunt for new and faster routes to the Montana gold camps. The Bozeman Trail is a good example. While speed was a factor, so was safety. Trails were filled with enough dangers without adding the threat of Indian attacks. One group of promoters in Minnesota believed they had a solution: the Northern Overland Route.

Between 1862 and 1866, Captain James Liberty Fisk led four expeditions from Minnesota across the northern Great Plains to Fort Benton on the upper Missouri River. Probably more than any other road, the trail made perfect sense. Yet it never really caught on with travelers until the later part of the nineteenth century, when James J. Hill built his St. Paul, Minneapolis and Manitoba Railroad (later the Great Northern Railway) along Fisk's old trail.

The first Fisk expedition in 1862 proved to be the most successful—at least, it was the most trouble free. The expedition included 117 men, 13 women, a military escort of 50 volunteer cavalrymen, a physician, oxen, mules, cows, horses and "numerous dogs." It also included 40 immigrants who enlisted as guards and scouts in return for their board on the long trek west. The outfit traveled directly west from near present-day Fargo, North Dakota, and reached Fort Union near the confluence of the Missouri and Yellowstone Rivers about a month after its departure. After a three-day layover, the expedition followed the Missouri River north to the confluence of the Milk River and then followed that course west across Assiniboine territory, reaching Fort Benton in early September.

At Fort Benton, the military escort disbanded, with some following Fisk on to Walla Walla, Washington, on the Mullan Road. News of a gold strike on Prickly Pear Creek southwest of present-day Helena caused most of the sojourners to leave the main group to head for the diggings. Fisk and the remnants of the expedition continued on to Walla Walla. The first Fisk expedition over the Northern Overland Route proved the route practical. After his return to the states, Fisk obtained additional federal funding and made plans for a second expedition the following year.

The next three Fisk overland expeditions were not as successful as the first. The Sioux uprising in Minnesota in 1862 caused many of the

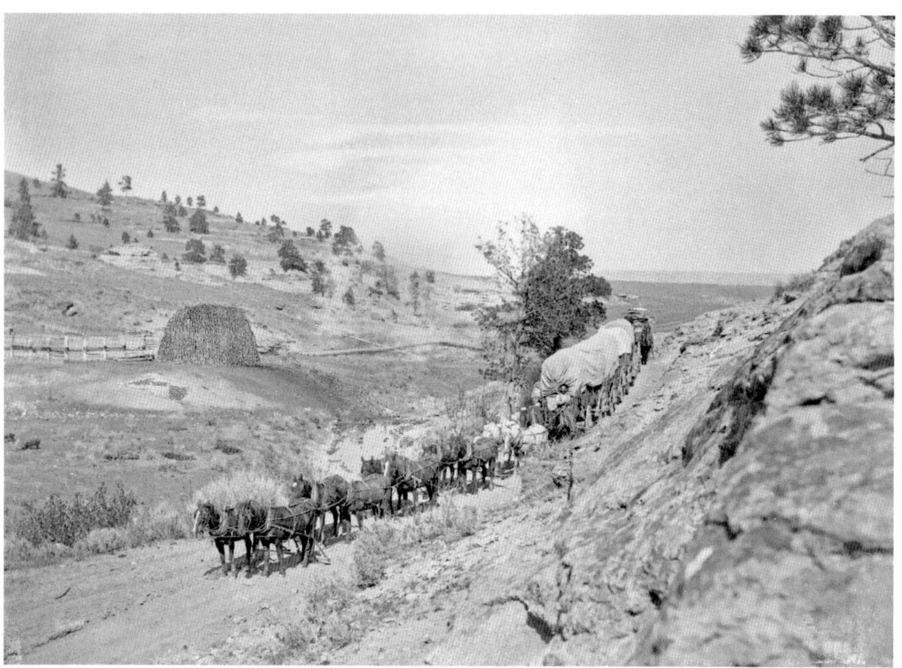

Freight wagons on the road between Coulson and Fort Benton in 1883. *MHS Photograph Archives, Helena, 981-248.*

Indians to flee to Dakota Territory. The Bozeman Trail further increased tensions with the Indians on the northern Great Plains. Fisk planned for confrontations with the Indians by improving the firearms of the soldiers and carrying a twelve-pound howitzer with the expeditions. He also explored different routes for the trail, some of which crossed through prime bison hunting grounds protected by the Lakota. A running battle with the Lakota in 1864 forced the third expedition to turn back to Fort Rice rather than risk further danger by continuing west. Fisk planned another expedition in 1865, but poor planning and his chaotic personal finances forced him to postpone the trip until 1866. The fourth and final Fisk expedition in 1866 included 160 wagons and 500 people, mostly merchants carrying their goods to the Montana mining camps. Although attacked by Indians, the expedition made it through to Montana, but this proved to be the last official trans-northern Great Plains wagon train to the territory.

Toll Roads

By the time Montana had become a territory in May 1864, the Montana and Mullan Roads were well-established supply lines to the mining camps. Connecting roads branched from these main routes to the mining camps, the Gallatin and Missoula Valleys and Fort Benton. The Organic Act that created Montana Territory in May 1864 made no provision for the taxation of Montana's residents to pay for roads and bridges. Indeed, the territory could not charge property taxes until the General Land Office surveyed the land and made it available for private acquisition under the 1862 Homestead Act. The first territorial legislature, therefore, needed to address the issue of how the territory could establish and maintain a decent road system with no tax base.

Meeting in Bannack from December 12, 1864, to February 9, 1865, the first territorial legislature was a raucous one. The twenty-two-man body was divided by Civil War sentiments, with the number of Republicans and Democrats evenly matched. The assembly was well aware of the importance of transportation to the prosperity of the new territory. Consequently, it passed a law that granted licenses or franchises to toll road, bridge and ferry companies. In January 1865, the territorial legislature granted charters to thirty-five toll companies. Located mostly in southwestern Montana, the companies built and maintained the roads and bridges that connected the mining camps to the Mullan and Montana Roads.

A tollkeeper on a well-established route could make thousands of dollars over the course of a season with only a minimal amount of work put into the road. Territorial laws required that the facilities be maintained but didn't specify how much work a tollkeeper needed to put into a road to keep it in a passable condition. Travelers' complaints to the territorial legislature about rickety, unsafe bridges and dreadful roads were common. The legislators had no way to enforce the law. There is no evidence that they revoked any toll licenses for non-maintenance of a road or bridge during the six years the toll system was in use in Montana Territory. The pattern in Montana suggests that if a road or bridge wasn't properly maintained, then the freighters and stagecoach companies would simply find a new route—and there were plenty of people willing to accommodate them. Most of the toll companies, however, knew where their bread and butter were coming from and maintained their facilities.

In 1865, the legislature granted a license to Ed Lewis and Malcolm Clarke for the Little Prickly Pear Wagon Road north of Helena. The road bypassed the more rugged Mullan Road, which was located a few miles to the west of Prickly Pear Canyon. The partners built a rudimentary road with twenty-seven fords of Little Prickly Pear Creek. Two years later, they sold the road to James King and Warren Gillette, who may have spent as much as $40,000 reconstructing the road. They eliminated all the fords, built nine bridges and

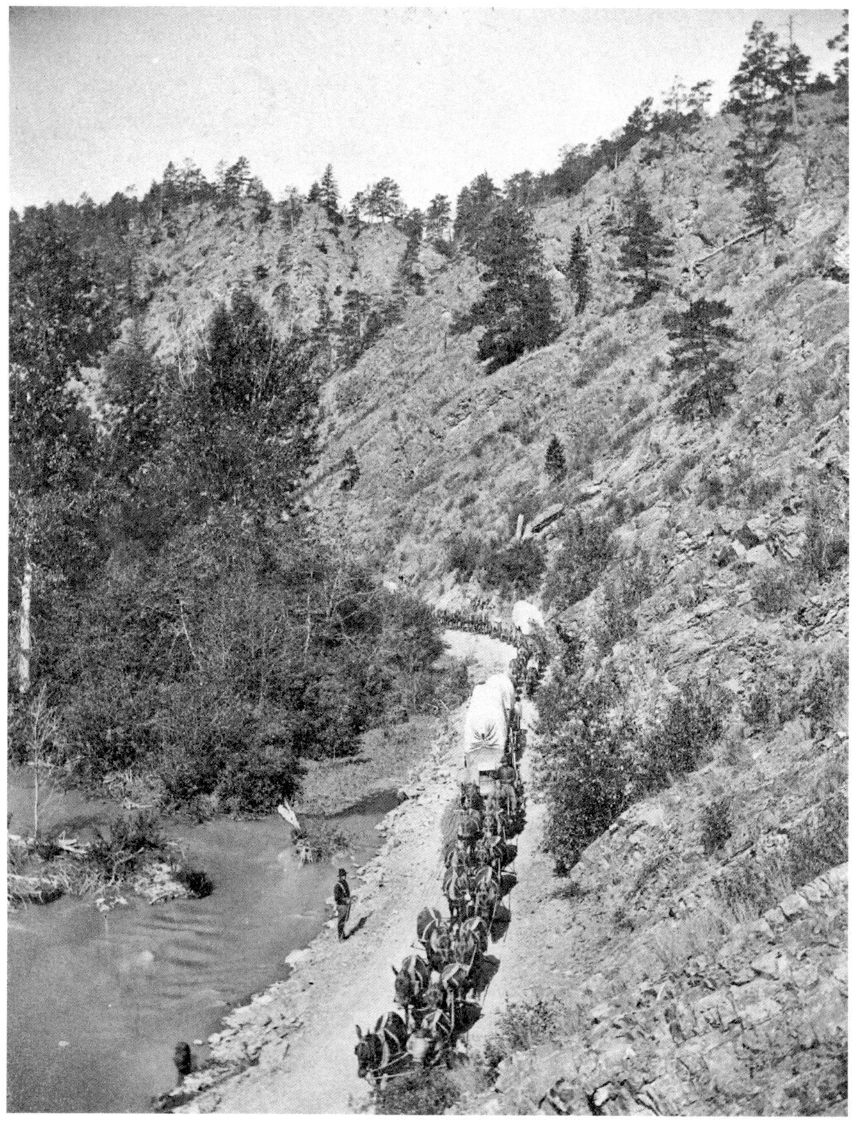

The Prickly Pear Canyon echoed with the sounds of Hugh Kirkendall's wagon trains in 1866. *MHS Photograph Archives, Helena, Stereograph Collection.*

dug road "cuts nearly throughout its whole extent." The *Helena Tri-Weekly Republican* reported about their road:

> *The cemetery of the giants was defiled and the castle walls frowned upon the desecration. A passage way was hewn through the rocks and trees and constructed across the ravines and bogs at the bottom of the defile, and man looked up triumphant where before he looked down and acknowledged his own impotence. Thanks to the energy of Messrs. King and Gillette...a wagon road had been built through a place, which it was supposed by many would forever remain impassable. This road is some six miles long and has, without doubt had more work done on it than has been bestowed upon the entire length of the traveled way between Helena and Salt Lake* [City].[7]

With the partners' improvements, traffic on the road substantially improved, making the investment a profitable one for King and Gillette. They recouped their $40,000 investment within two years!

Other toll roads succeeded more as stagecoach routes than as freight roads. In December 1866, the territorial legislature granted a license to French Canadian Constant Guyot to build a toll road over the Continental Divide west of Helena. The legislators awarded the license despite the opposition of territorial governor Green Clay Smith. Smith had problems with the toll system already but was mortified that the legislators would allow Guyot to collect tolls on a road that hadn't been built yet. The road, when it opened in 1867, provided a more direct route across the divide from Missoula and Deer Lodge than the older Mullan Pass route. An advertisement in the *Deer Lodge Weekly Independent* advertised Guyot's toll road as "thoroughly STAKED OUT, so that it will be impossible to go astray while the snow is on. Travelers can be accommodated with meals and lodgings at the French Woman's." The "French Woman," Madame Guyot, was one of many female toll collectors in Montana; the road carried her name until well after her tragic death in 1868.[8]

Details about Madame Guyot are meager, including her first name. She is referred to in the historical record simply as the "Frenchwoman" or the "Old Frenchwoman." No photographs of her are known to exist, and few written descriptions of her are available. In 1881, the *Fort Benton River Press* described her as a "neat looking critter—black-haired, black-eyed, and sharp, and cute lookin', maybe thirty years old, an' a good housekeeper." A more contemporary description described her as a "garrulous, gossiping, good natured dispenser of ranch eggs and trout and tortured English."

While accounts praised Mrs. Guyot, the reports of her husband are less than flattering: he was hard drinking and abusive.[9]

The Frenchwoman maintained a hotel in a log cabin near Dog Creek at the west end of the road near the junction of the toll road and the Mullan Pass road, about eighteen miles west of Helena. One account states that as many as thirty men could be found sleeping on the floor of her two-room "hotel" at any one time. Lodging at the Frenchwoman's was two dollars a night, and meals were one dollar, served in the same room where people slept. All transactions utilized gold dust. Constant Guyot spent most of his time working a hay ranch about two miles away; he had little to do with the day-to-day operation of the toll road and roadhouse.

A large number of the Frenchwoman's customers were stagecoach passengers. Wells Fargo and Company designated the road as part of its route between Deer Lodge and Helena in 1867. Stagecoaches followed the gold miners into Montana Territory in 1863. Oxen and mule trains may have carried freight to Montana, but the stagecoaches carried the equally important mail and passengers to the territory. A network of stage stations crisscrossed western Montana by 1866. There is little evidence that the stage companies did much work improving roads, instead relying on the toll operators and the counties to do the work. It is not known what arrangement the stage companies had with the toll collectors, but stagecoaches shared the toll roads with the freighters. Stagecoaches were a necessary evil—a tough way to travel suitable only for the hardiest souls.

Typically, stagecoaches held nine people inside the cabin with as many as twelve people sitting on top of the coach in addition to the driver and conductor. Stagecoaches also carried mail and baggage. Since the stage companies were in the business to make money, it behooved them to carry as much mail and convey as many passengers as possible. Each wagon carried as much as three tons and was drawn by six to eight horses. Changing stations were located every ten to fifteen miles, while the home stations, where passengers could get out and stretch their legs, relieve themselves and get a meal, were located every forty to fifty miles. Travel time was based on distance, road conditions, the weather and, occasionally, whether or not the driver was drunk. It was not uncommon for stagecoaches to overturn, suffer mechanical breakdowns or become mired in mud or stuck in the snow. The Frenchwoman's functioned as a home station.

In 1868, travelers seeking breakfast found Madame Guyot murdered in her place of business. Although nobody ever stood trial for the crime, evidence strongly pointed toward her husband, who had prudently left the

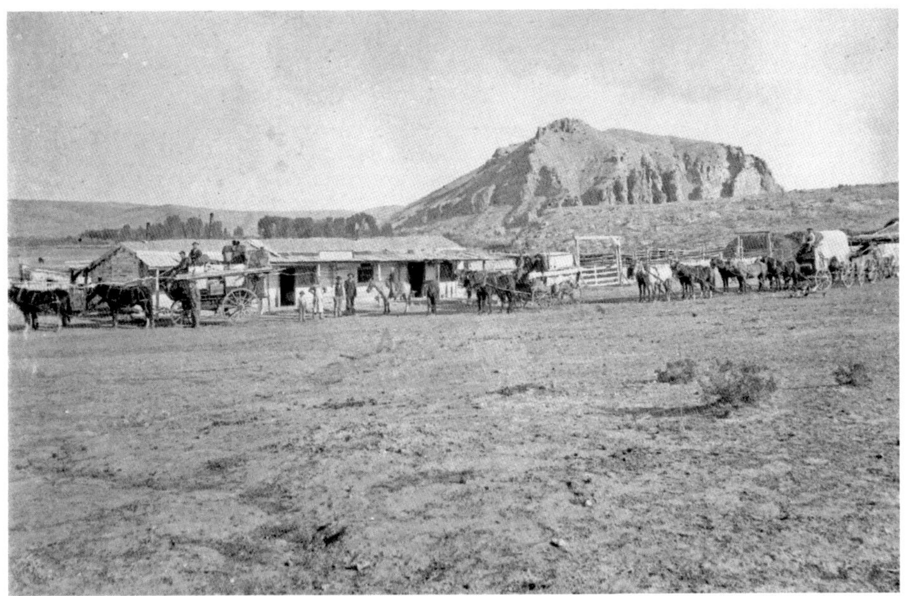

A stage station on the road between Bannack and Virginia City, near Beaverhead Rock. *MHS Photograph Archives, Helena, 952-891.*

territory. Elijah Dunphy, an employee of the Guyots', obtained the license and continued to operate the toll road until Lewis and Clark County took it over in the 1890s. The pass over the Continental Divide today is named after one of Dunphy's hired hands, Alexander MacDonald.

For six years, travelers in Montana endured the torment of navigating a maze of toll roads and bridges. While the legislators may have established the rates toll companies could charge travelers, evidence suggests that price gouging was a common occurrence on the territory's roads. During his address to the legislature in 1869, territorial governor James Ashley called the current toll system "little better than legalized highway robbery" and complained that "travelers and freighters found, in every canyon, on almost every water course and on many broad and level plains, a toll collector who demanded, as a condition to the passing of each, from one to three dollars." A team with a loaded wagon could expect to pay $40 in tolls between Helena and Corinne, Utah. That would roughly translate to $570 in early twenty-first-century dollars. Stagecoach operator and freighter Herman Reinhart wrote that on one trip between Helena and Virginia City, he "had so many roads and ferries to cross...that it took off a good deal of the profits of the trip." No one was free from the tyranny of the toll collectors; even the U.S.

Stagecoach Etiquette

Never ride in cold weather with tight boots or shoes or close-fitting gloves. Bathe your feet in cold water before starting, and wear loose overshoes and gloves two or three sizes too large. When the driver asks you to get off and walk, do it without grumbling. He will not request it unless absolutely necessary. If a team runs away, sit still and take your chances; if you jump, nine times out of ten you will be hurt. In very cold weather, abstain entirely from liquor while on the road; a man will freeze twice as quick while under its influence. Don't growl at food at stations; stage companies generally provide the best they can. Don't keep the stage waiting; many a virtuous man has lost his character by so doing. Don't smoke a strong pipe inside, especially early in the morning. Spit on the leeward side of the coach. If you have anything to take in a bottle, pass it around; a man who drinks by himself in such a case is lost to all human feeling. Provide stimulants before starting; ranch whiskey is not always nectar. Be sure and take two heavy blankets with you; you will need them. Don't swear or lop over on your neighbor when sleeping. Take small change to pay expenses. Never attempt to fire a fun or pistol while on the road; it may frighten the team, and the careless handling and cocking of the weapon makes people nervous. Don't discuss politics or religion. Do not point out places on the road where horrible murders have been committed if delicate women are among the passengers. Don't linger too long at the pewter washbasin at the station. Don't grease your hair before starting or dust will stick there in sufficient quantities to make a respectable "tater" patch.[10]

Army had to pay a fee for each soldier and each of its wagons. The freighters passed on the toll costs to the territory's consumers, which drove up prices and impeded Montana's economic growth. The problems eventually forced the territorial legislature to deal with the grossly inefficient and unpopular toll system.[11]

When the Fifth Territorial Legislature met in Virginia City in December 1870, it significantly changed the system. After Governor Ashley's impassioned condemnation of the toll road system, the legislature repealed all of the toll licenses. Instead, it made the county commissioners responsible for maintaining the territory's road system. The legislature directed the commissioners to establish road districts and appoint road supervisors to maintain them. To pay for the roads, the counties levied a road tax on property owners. The counties also charged a poll tax of three dollars to help maintain the roads. Citizens unable to pay the tax could work it off on road maintenance. The legislature allowed the counties to charter toll roads and bridges. By 1883, there were only four known toll roads in Montana: Yankee Jim's National Park Toll Road in the Paradise Valley south of Livingston, Browne's Bridge northeast of Bannack and the Priest's Pass and MacDonald Pass roads over the Continental Divide west of Helena. There were no toll roads in Montana by the second decade of the twentieth century.

The Iron Horse Transforms Montana's Roads

The arrival of the railroads in the 1880s had a profound impact on Montana. The railroads, more than gold or silver, were responsible for the settlement of the territory. Wherever they went, they deposited towns in their wakes and brought thousands of new settlers to the territory. The cities lucky enough to be on the lines experienced economic booms and significant growth. The railroads hastened the slow decline of Virginia City and Bannack, which were not on lines. They allowed investment in the territory by eastern capitalists. Consequently, the copper and silver mines in and around Butte boomed. The railroads sparked the establishment of the great cattle empires in eastern Montana as they provided direct access from the Great Plains to the packinghouses in Chicago. Montana copper and cattle were critical to America's industrial revolution.

The railroads significantly changed transportation in Montana as well. No longer did travelers have to rely so much on cramped stagecoaches, and the wagon trains quickly disappeared. The railroad's ease of travel and ability to carry more freight than wagons quickly rendered the old ways of getting around the territory obsolete. Instead of the two days it took to travel to Fort Benton from Helena, it took only a few hours on the Montana Central Railroad after 1887. Likewise, supplies shipped north from Utah arrived in Butte after only days instead of months.

Other changes to the transportation landscape also took shape beginning in the 1880s. Significantly, roads began to serve the railroads instead of the settlements. Many railroads, like the Montana Central through Prickly Pear

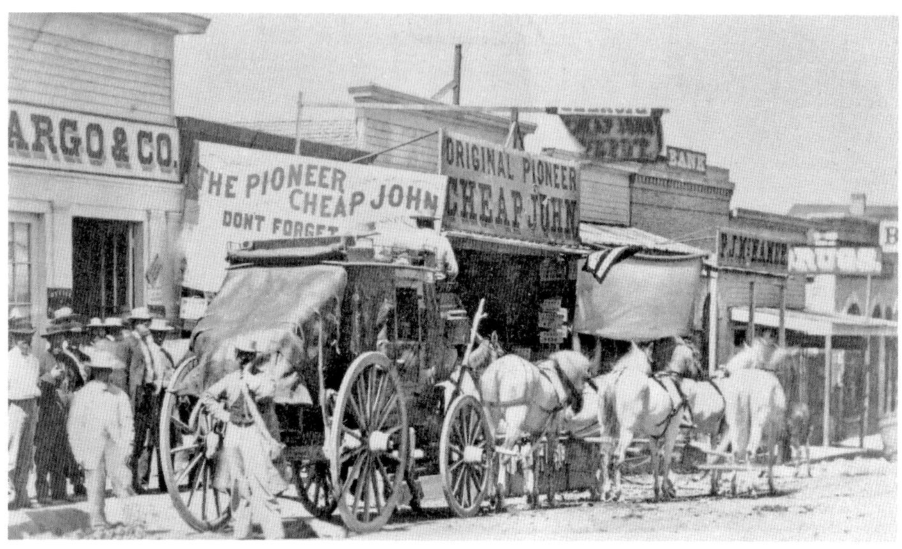

A stagecoach readies for departure at Helena's Wells Fargo office, 1866. *MHS Photograph Archives, Helena, 952-949.*

Canyon, occupied the routes of former roads. Importantly, the railroad allowed the importation of steel and iron for bridge construction. The significance of this manifested in the near disappearance of river ferries, replaced by the more reliable steel and iron bridges in the late nineteenth century. The first of these was the Missouri River Bridge at Fort Benton.

With the end of the steamboat era in the early 1880s and the arrival of the St. Paul, Minneapolis & Manitoba Railroad in 1887, Fort Benton aggressively sought to maintain its preeminent economic position in Montana Territory. The railroad provided direct access to national markets through Minnesota and a connection to the flourishing mining camp of Butte to the south. It was also strategically perched on the edge of one of the most productive cattle regions in the territory: the Judith Basin. The question for Fort Benton's businessmen was how to take advantage of the economic opportunities offered by the Judith Basin. Ferries were inefficient and seasonal; not enough supplies could be provided to the area by ferry, and swimming cattle across the river at that point wasn't feasible. The city's businessmen hit on the idea of building a steel bridge across the Missouri, thus providing an all-season river crossing to the Judith Basin. To that end, they collected money from area businesses and got the Chouteau County commissioners to bless the project. Milwaukee Bridge & Iron Works began construction of the bridge in February 1888 and completed it in December of that year.

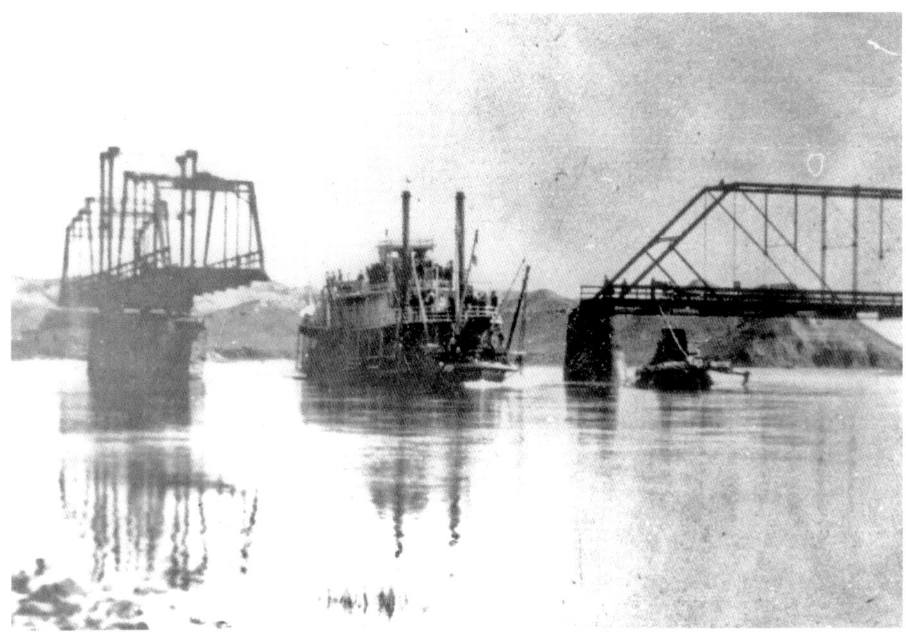

The steamboat *OK* and the Fort Benton Bridge in 1908. *Fort Benton Library.*

The bridge, which still stands, was the first all-metal bridge built in Montana and opened the Judith Basin to the railroad. Because the federal government considered the Missouri River navigable to Fort Benton, the bridge's design included a swing span to allow the passage of steamboat and barge traffic. The bridge accomplished what its builders intended, and Fort Benton prospered during the heady days of the cattle boom. The bridge opened the door for other counties to erect steel bridges in their jurisdictions. The railroad provided direct access to eastern United States bridge factories, which could ship the dissembled structures to Montana, where they were erected in great numbers. Most were simple steel truss bridges that were held together by steel pins. The pins allowed the manufacturer to build a bridge to the county's specification, take it apart, ship it west and reassemble it with relative ease. Pin-connected bridges were common sights throughout Montana by the second decade of the twentieth century. There were also variations of the design, such as the unique Dearborn River High Bridge southwest of Augusta and the Clark Fork Bridge at Thompson Falls. Steel bridges have been the hallmark of modern road building in Montana since 1888.

Part II

GETTING MONTANA OUT OF THE MUD

In no state in the Union has the automobile increased in the number of devotees in the last year than in Montana. Also probably true in no other state has there been more attention given to the matter of road improvement. In every one of the larger cities and towns of the state there are enthusiasts banded together for the one purpose of better roads. Next to railroad development the improvement of the highways is regarded as one of the first importance keeping Montana the treasure state of the Union.

—Anaconda Standard, *December 14, 1913*

Montana experienced profound changes at the beginning of the twentieth century. The Enlarged Homestead Act of 1909 opened up twenty-five million acres of public and railroad-owned land to homesteading. Well over half a million people moved to the state and filed on homesteads within just a few short years. The massive influx of new people into eastern Montana tipped the economic balance of the state to farming, replacing mining as the state's primary source of revenue. With the thousands of new residents came new towns founded to serve the needs of the homesteaders and their families. The new residents clamored for better and more easily accessed government services, which resulted in the doubling of the number of Montana counties from twenty-eight in 1910 to fifty-six by 1925. The homestead boom coincided with above-average rain amounts, producing bountiful wheat harvests. The war in Europe caused prices to soar for American agricultural products. Mining and smelting in Butte and

Anaconda also boomed as the European war caused copper prices to rise dramatically. The booming economy and thousands of new residents placed gargantuan demands for better roads on county governments. The Montana Good Roads movement peaked in connection to the homestead and mining booms. For a time, the counties had the financial resources to meet the demands of their constituents. Because of the demands of Montana's new residents, the state legislature focused much attention on the state's road problems and aggressively sought ways to solve them.

It is not known precisely when the first automobile appeared in Montana. But by 1913, there were over 6,000 vehicles registered to Montanans. That number had jumped to over 14,400 vehicles by 1916. The usage of most were confined to cities and towns; long-range trips in them were rare, an endeavor for only the most adventurous. The most serious obstacle to motorists in the early twentieth century was the lack of good roads. At the dawn of the automobile age, Montana's roads weren't much better than they were in the 1860s. County governments endeavored to construct as much mileage as possible with the limited funds at their disposal. Unfortunately, in the counties' haste to construct roads, the resulting

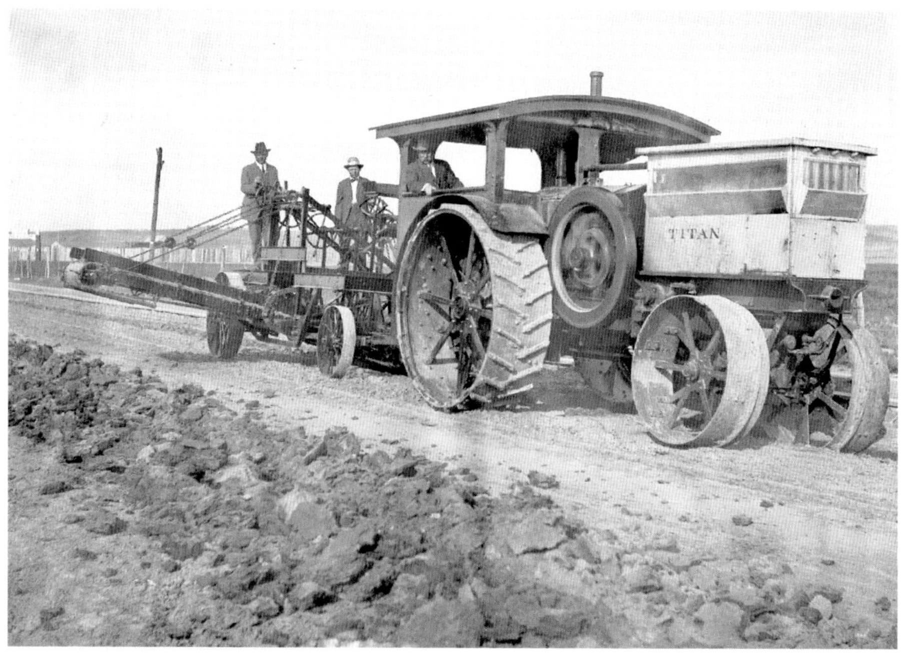

A Titan barrow pit tractor and conveyor, 1912. *MHS Photograph Archives, Helena, 957-321.*

thoroughfares often lacked any real durability and were difficult to maintain. Primarily earth-surfaced roads, they were gumbo morasses after the rain and during thaws, and they were badly rutted and washboarded in the summer.

While their road systems may have been abysmal, the counties were active bridge builders. They built steel truss bridges at major river and stream crossings and steel stringer, timber and reinforced concrete structures at minor crossings. By the second decade of the twentieth century, the counties had systemized the process of constructing bridges. County commissioners provided basic specifications, while the many bridge construction companies active in the state built the structures. The industry, however, was unregulated, with little oversight of the companies by the counties. The counties sometimes found themselves owning bridges that were overdesigned, badly constructed or totally unsuitable to their needs. Critics, including professional bridge engineers, clamored for some level of state or federal control of the bridge-building industry.

Before 1913, there were at least thirty bridge construction companies active in Montana. While a few were based in the state, most were from the Midwest, which had direct access to Montana over the railroads. Bridge construction was a lucrative business, with an overabundance of companies competing in a limited market. Beginning in the late 1890s, the companies resorted to a system originally developed by the railroads to guarantee them business: pooling. Pools were unofficial agreements among the companies establishing specific areas where they would be guaranteed work. On average, seven to ten bridge construction companies submitted bids for county bridge projects. According to the pool agreements, one company would consistently submit the low bid in a specific area and win the contract for the bridge. The Billings-based Security Bridge Company almost always got the contracts for bridge projects in central and south-central Montana. The O.E. Peppard Company of Missoula got all the bridge jobs in western Montana and along the High Line. The King Bridge Company of Cleveland was often the successful bidder in Lewis and Clark County, while the Gillette-Herzog Manufacturing Company was the favored bridge contractor in northwestern Montana. Although pooling worked well for the construction companies, it was an illegal system that eventually drew the attention of the state legislature.

THE MONTANA STATE HIGHWAY COMMISSION

Federal, state and public agitation for better roads began in the late nineteenth century and peaked in the 1910s. State lawmakers and Good Roads advocates supported efforts by the federal government to enact legislation that would give it and state governments greater roles in the improvement of postal delivery routes. The Montana legislature failed to pass a bill to create a state highway law in 1911 but tackled it again two years later. In March 1913, the Thirteenth Montana State Legislature unanimously passed a bill creating the Montana State Highway Commission. The new highway commission marked the culmination of efforts by Good Roads enthusiasts, farmers and motorists in the state to establish some form of centralized control over road building in Montana. The formation of the highway commission coincided with the enactment of the first Motor Vehicle Law, which required all vehicle owners to register their vehicles with the secretary of state. The state and counties used funds obtained by vehicle licensing to raise "revenue for the constructing, maintenance, and improvements of public highways."[12]

The first highway commissioners were civil engineers. Governor Sam Stewart appointed Robert D. Kneale, Archibald W. Mahon and George R. Metlen to the commission in April 1913. Kneale was a professor of engineering at the Montana State Agricultural College in Bozeman, while Mahon had been the state engineer since 1911. Metlen, a civil engineer from Beaverhead County, served as the commission's first secretary and was the designated contact person with the counties. He was also the only member of the commission to receive a salary: $3,500 per year. The legislature allocated $5,000 for the administration of the highway commission; it had no resources to design or build roads.

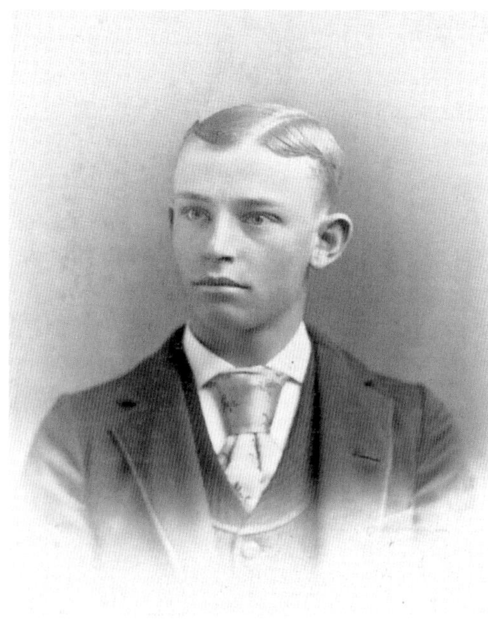

George Metlen. *Beaverhead County Courthouse, Dillon.*

Under the terms of the legislation that created the commission, it could only "give [the counties] such advice, assistance, and supervision with regard to the road construction, improvement, and maintenance throughout the state as time and conditions would permit." The law encouraged the counties to work with the new highway commission but did not make it mandatory that they do so. The commission made recommendations for new roads, developed standards for their construction and identified material sources in the state. It established rules and regulations regarding materials for road construction and published pamphlets on the best practices for the use of road machinery, surfacing materials and drainage. Although the legislature intended the commission as a centralized authority to oversee the development of the state's road system, it, in fact, had no real authority over the counties.[13]

Initially, the counties were responsible for road surveys, the establishment of road grades, the preparation of plans and preliminary estimates for all work on designated state roads within their jurisdictions. The highway commission reviewed the counties' plans and specifications and could make recommendations, but the counties were not obligated to accept them. The counties were also responsible for advertising the projects and awarding contracts to the contractors. It was only over the convict road crews that the highway commissioners had any authority.

At the first highway commission meeting in April 1913, Kneale and Mahon directed Metlen to investigate the possibility of using convict labor on the construction of state roads. In Montana, state penitentiary warden Frank Conley had used prison convicts to build roads in the Deer Lodge area since 1910. Conley believed that physical labor was a means of rehabilitating prisoners by creating a structured environment and allowing them some responsibilities outside the prison walls. Conley's program impressed Metlen, who reported favorably on the use of such labor at the commission's next meeting. Metlen reached an agreement with the state prison board on behalf of the highway commission to oversee the use of convict labor on state highway projects in the summer of 1913.

Under the provisions of the agreement, the state prison board provided convicts to the counties for projects approved by the highway commission and Conley. The counties paid fifty cents per day per convict for upkeep, while the prison provided the guards. The counties were also responsible for the transportation of the convicts to and from the prison to the job site. The highway commission supervised the work and provided equipment to the convict work crews. The first road built by convict labor under the direction of the highway commission was on the east side of Flathead

Lake south of Bigfork. By 1915, convict crews built 119 miles of road in seven counties. Early on, however, Metlen recognized that the convict road crews were not cost-effective for light work but were profitable only on projects involving heavy rock excavation and timber clearing. Prison crews were the only experienced road builders in the state when World War I ended.

In May 1915, the Montana legislature passed legislation that required the highway commission to form a bridge department. The new department, under the direction of Charles A. Kyle, developed standard plans for steel truss, concrete and timber structures. It also developed bid packages for contractors and provided a supervisor at the job sites. The counties remained responsible for financing the structures, which could be built only on state-designated highways. While not all counties chose to follow the new regulations, the majority of them did, which, in turn, put controls on the bridge companies and further promoted a standardized Montana highway system. During the first year of the program, Kyle supervised the construction of sixty-nine bridges in 1915 alone.

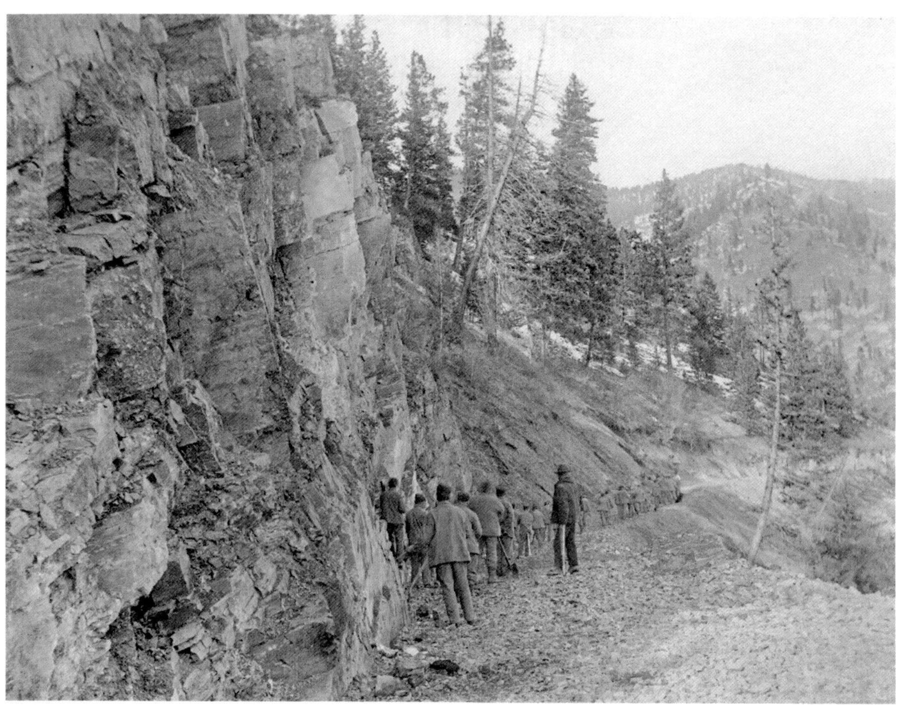

A convict road crew on the Yellowstone Trail near Nimrod, 1921. *MHS Photograph Archives, Helena, 949-985.*

Counties' efforts to improve their roads continued. In 1916, a large federally funded project occurred in Ravalli and Beaverhead Counties between Sula and Wisdom. The Trail Creek section of the Big Hole Road was located in the Bitterroot National Forest. That meant the road was designed and funded entirely by the U.S. Department of Agriculture (USDA). In 1914, the USDA hired the Spokane, Washington–based firm of Clifton, Applegate and Toole to construct fifteen miles of the road, with Big Hole ranchers constructing the last three miles of the project. The contractor began work on the project in January 1915. The project began after USDA and county surveyors established the alignment of the road. They were followed by workers who cleared all the trees and brush from within the projected right of way. The foreman marked all trees to be cut down, and men called "swampers" felled the trees and cut them into logs. They stacked the logs next to the road along with the trimmed branches and brush. A newspaper reporter observed that the "logs and brush are piled so thick that it is almost impossible to get a saddle horse off the road."[14]

Because the proposed route would be located on mountainsides, significant grading was necessary to build the road. Work on the mountainsides required the use of dynamite to blast through the rocks to obtain the necessary 6 percent grades. In areas where blasting was not necessary, a large railroad plow dug up the ground, followed by a grader that smoothed the road. Ten horses pulled the grader, which caused a dust cloud that could be seen over a mile away. A newspaper reporter added that "the dust was usually tinged with blue from the remarks made by the 'skinners' and 'muckers' who worked with the grader, as it was one of the worst jobs on the line." When the grader had finished its work, the "finishing gang" put the last touches on the road. The contractor built all bridges and culverts from logs acquired in the nearby forest.[15]

The project's workers were a mix of Americans and immigrants. The unskilled labor consisted mostly of Italians and Eastern Europeans. They did the pick and shovel work while Swedes did the clearing and bridgework. The teamsters, for the most part, were Americans and Irishmen. The newspaperman reported that the Eastern Europeans were the easiest to handle, unless they decided they didn't want to do the work. The American-born workers, the author wrote, were complacent and followed the foremen's orders. All the workers lived in camps adjacent to the construction zone. Because of the remoteness of the project, the contractor sometimes had difficulty getting supplies to the camps. The project demonstrated that the

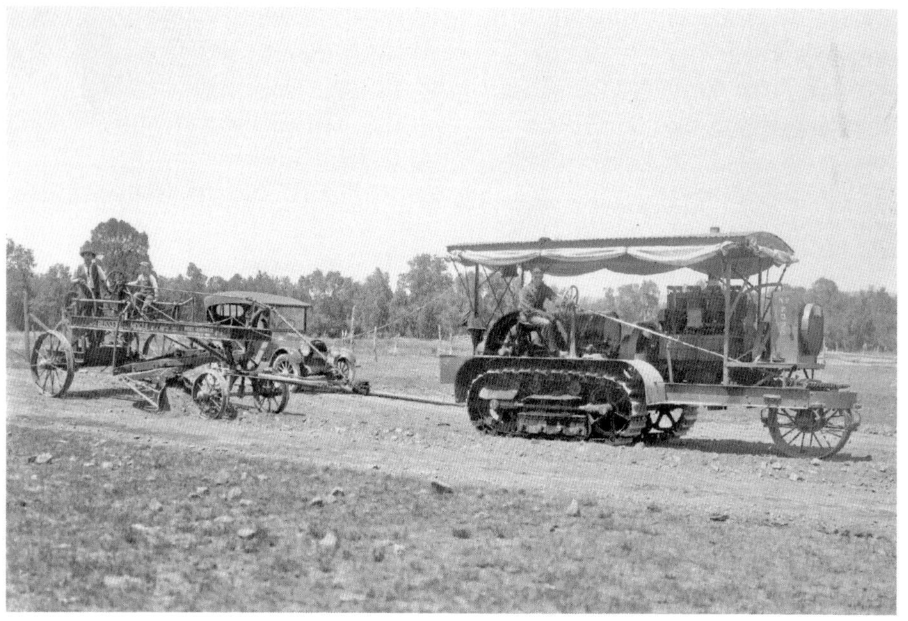

A tractor towing a grader at Bridger Creek in Sweet Grass County, 1916. *MHS Photograph Archives, Helena, 957-342.*

federal government had more control over road construction in Montana than the state's highway commissioners.

Although the highway commissioners took on their responsibility in 1913 with great hopes for the future of road building in Montana, their optimism was tempered by the reality of the situation. The legislature was reluctant to centralize too much authority over road building with the commission in Helena. Some counties chose not to cooperate with the commissioners and were openly hostile to them. The commissioners themselves realized they could do little to establish a standardized highway system in the state as long as the counties controlled the finances. The 1913 law, moreover, did not clearly identify the commission's exact duties and authority. Other than the legislature's passage of a new bridge law in 1915, the commission was unable to make any meaningful improvements to the situation. Commissioner George Metlen concluded that in Montana, "with its rapid growth and settlement of heretofore undeveloped and new country, the demand for roads is coming faster than the money with which to pay their cost of construction of the cheapest character." The counties were more interested in quantity than quality. Fortunately, some salvation was at hand from the federal government.

The Federal Aid Road Act of 1916 had sweeping ramifications for the Montana Highway Commission and the counties. The legislation appropriated $75.0 million to the states over a five-year period; Montana received $1.5 million of that money. The state highway commission allocated the money to the counties on a fifty-fifty match basis. The commission was responsible for completing all surveys, plans, specifications and estimates for road and bridge projects, which would also need the approval of the federal Bureau of Public Roads (BPR) before the funds could be spent on projects. The highway commission could make payments to the counties only as the work progressed, with no payments to exceed $10,000. The BPR and the highway commission required the counties to maintain all roads constructed with federal money. The act also allocated money for construction of roads in federally administered national forests. Forest highway funds did not require matching from the counties in which the projects were located; they were constructed under the direct administration of the BPR.

Soon after the passage of the act in July 1916, George Metlen traveled to Washington, D.C., with other state highway commission members for a briefing on the legislation by the BPR. Metlen returned to Montana in early September and reported the information to his colleagues on the commission. That initiated three months of sometimes-heated debate among the commissioners about how to implement the new law. There is no record that the counties were included in the discussions or that Metlen sought their opinions. It became obvious to lawmakers that the commission, in its present form, would be unable to administer the federal legislation. The law required a proactive highway commission with authority, not one that could only collect and disseminate data. The 1917 Montana legislature remedied that situation.

A New Direction

Although the formation of the highway commission was met with almost universal praise by Montana's counties in 1913, four years later, the public often condemned it for not doing enough—especially in the area of funding—to assist the counties in their road-building efforts. It was a problem of which the highway commissioners were keenly aware, but state legislation tied their hands. The highway commission recommended that the legislators revise the law to be more specific about

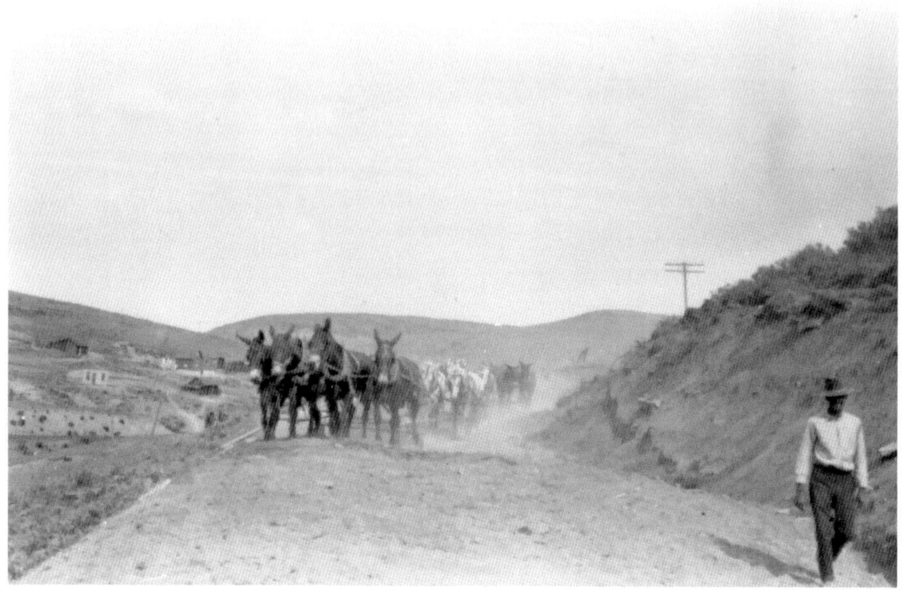

Building the road between Bearcreek and Red Lodge in Carbon County, 1919. The Smith coal mine is in the background. *MDT.*

its duties and authority. It stressed its need for control of the finances rather than giving it to the counties. The 1916 Road Act emphasized the importance of state highway commissions in the preconstruction and construction phases of road projects. As it was organized, the highway commission would be unable to meet the increased demands placed on it by federal and county governments. In March 1917, the state legislature reorganized the highway commission to better handle the demands of the 1916 federal law. The reorganization was a milestone in the development of Montana's road program and had significant engineering and political ramifications for the state for many years.

The new highway commission consisted of twelve men, one from each of the commission's twelve construction districts. The commissioners selected three men to form an executive committee composed of a salaried president and two assistants, who received only per diem and expenses for their time. The twelve-man highway commission met twice a year while the executive committee met monthly. The committee made all policy decisions, provided advice and assistance to the counties and oversaw all contracts for work on state highways. The other nine members provided only support to the executive committee.

In order to fund the commission, the legislature raised the Motor Vehicle Tax in 1917 from two dollars to ten dollars per vehicle. The state allocated 75 percent of the revenue derived from the tax to the highway commission, while the remaining 25 percent went directly to the counties to provide matching funds for federal allocations. The highway commission used its portion of the tax to organize a highway department. Because the "lack of centralized authority in a single executive officer led to confusion and uncertainty," the commission hired Paul Pratt as the first chief highway engineer in April 1917. For a time, Pratt and the chief bridge engineer co-managed the highway department. By 1918, however, sole responsibility for the management of the department had fallen to Pratt, who answered directly to the highway commission's executive committee.

After several conferences with county commissioners, county surveyors and professional road builders, the highway commissioners structured the highway department to best serve its increased federal and state responsibilities. In July 1917, the department increased the number of engineers on staff and hired district supervisors to "efficiently carry out the policies of the commission." Indeed, through the remaining years of the 1910s, the number of highway department employees steadily increased in direct proportion to the number of projects it designed. By January 1920, the highway department had 156 employees, including 85 field men.

The highway department became active in April 1917, the same month the United States entered World War I. The war caused labor and material shortages. Construction costs increased, and the few highway contractors in the state did not adapt well to the situation. The war coincided with the collapse of the homestead boom, making it difficult for the counties to sell bonds to raise matching funds. The highway department could not attract trained engineers, and those it did hire often left the department as the demands of the war or better economic opportunities lured them away. The BPR withheld approval of proposed projects and draft road plans because of the war. Chief Engineer Pratt directed all highway department employees, especially the field men, to be on the lookout for German saboteurs.

The war postponed the highway commission's plans but gave it an opportunity to develop policies and reflect on the problems it would face when the war ended. By far, the biggest challenge was the system itself, specifically the counties' responsibility for the funding of projects on Federal Aid highways. The method was economically unreliable and politically fractious. Often, the counties proposed a project, on which the commission expended considerable money, and then withdrew from it, leaving the commission having spent money that could have been better spent elsewhere. The problem of selling bonds increased as the war progressed, leaving the counties without matching funds for Federal Aid. The highway commission's lack of direction prior to 1917 also caused its share of problems.

Most frustrating to the highway commissioners was the problem of selling projects to counties that would not directly benefit from the roads. The new commission's first biennial report to the legislature in 1919 stated:

> *In the development of a system of main state highways to connect trade and population centers of the state it is found that in many instances connecting links are required that are of very little local interest or advantage and that have little source of local support. Such highways often run through some mountainous section of a county with no local industry or source of revenue. Sometimes they run across some remote corner of the county while the source of highway revenue and the highway requirements of the taxpayers of the county are in a wholly different direction.*[16]

The commission understood that the construction of some connecting highways placed unnecessary financial and maintenance burdens on counties for which they would realize little benefit. The highway and county commissioners knew that as long as the system of funding and construction existed, this would continue to be a problem that would impede the development of the state's primary highway system.

During the war, the highway commissioners, Pratt and the BPR developed the basic system by which road and bridge projects would be constructed until the late 1920s. The highway commissioners prioritized and initiated projects in consultation with the county commissioners. The highway department developed projects in consultation with the BPR. When the BPR approved the project, the department assigned a survey party to establish the route of the highway and develop the plans based on the surveyors' findings. Once the BPR approved the plans, the highway department's engineers finalized them and directed the county commissioners to advertise the project and oversaw the awarding of the project to a contractor. The department placed a "resident engineer" on construction projects to ensure that the contractor followed the plans.

Although the role of the highway department in the program became more centralized because of federal funding, the counties still played a critical role in the process. They were responsible for acquiring the rights of

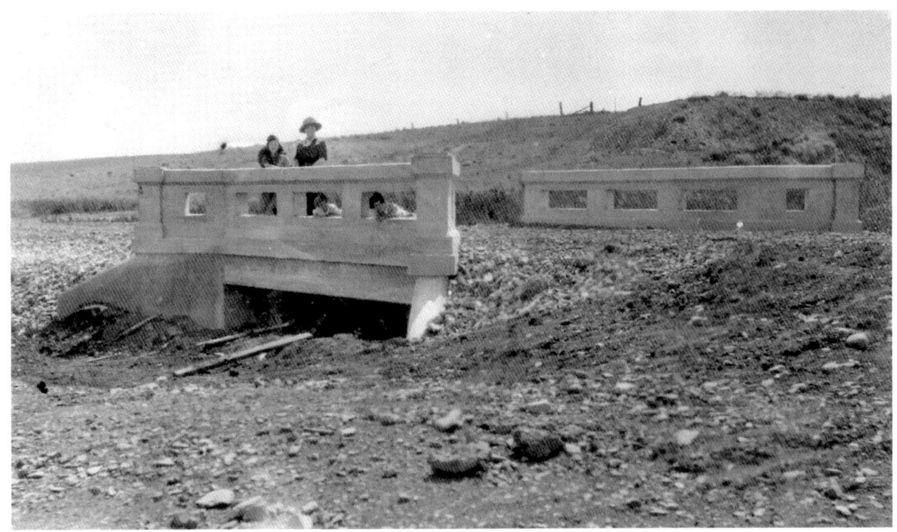

A reinforced concrete bridge near Roy in Fergus County, 1920. *MDT.*

way and paid half the project construction costs. The highway department was responsible for surveying, designing and supervising the construction of roads and bridges. The BPR set the standards by which the state and counties operated. Consequently, complaints by the counties surfaced about how the BPR and state highway department were bankrupting them to pay for "too expensive projects." That led to an unsuccessful attempt in 1919 by several Montana state senators to abolish the twelve-man commission in favor of a three-man highway commission. The rift between the highway department and the counties widened in the following years, especially as economic depression, drought and dwindling tax revenues crippled the counties after the war.

Ironically, the "ramped up" highway commission program in 1919 coincided with the collapse of the homestead boom caused by a severe drought and post–World War I economic depression. It was, according to many historians, "the most calamitous year Montana ever saw." Drought, which began in northern Montana in 1917, had spread over much of the state by 1919. That, coupled with low prices for agricultural goods in the wake of the war and reduced demand for copper, caused the state's economy to collapse. Drought and depression forced over 220,000 people, approximately 29 percent of the state's population, to flee Montana. The depression, unemployment and a diminishing tax base caused significant problems for county road projects; they lacked the revenue to provide matching funds for federal highway money. The highway commission understood the problems posed by the counties but was hampered by the provisions of the 1916 Federal Aid Road Act. In an attempt to alleviate the problem, the 1919 legislature passed a bill that made it easier for the counties to issue road construction bonds. The highway commission passed a resolution to assist the counties in their efforts to raise money but also resolved to recoup state funds lost because of failed bonds.

In April 1919, the first executive committee president's term expired, and the commission replaced him with Montana State Penitentiary warden Frank Conley, who had long experience with road building. At the time of Conley's appointment to the executive committee, his convict crews were excavating rock on an eight-mile section of the Yellowstone Trail near Nimrod in Granite County. The $170,000 project employed thirty-three men and earned the state penitentiary around $50,000. Convict crews were paid by the counties in which they worked, but the highway commission supplied the road construction machinery. Conley was a dynamic and forceful leader because of his long career as a prison warden and his experience in road construction.

The federal government aided substantially in the department's acquisition of construction machinery. At the end of World War I, the U.S. government was the largest owner of trucks in the world. Congress directed the secretary of war to transfer to the secretary of agriculture all vehicles, construction equipment and supplies "not needed by the military but suitable

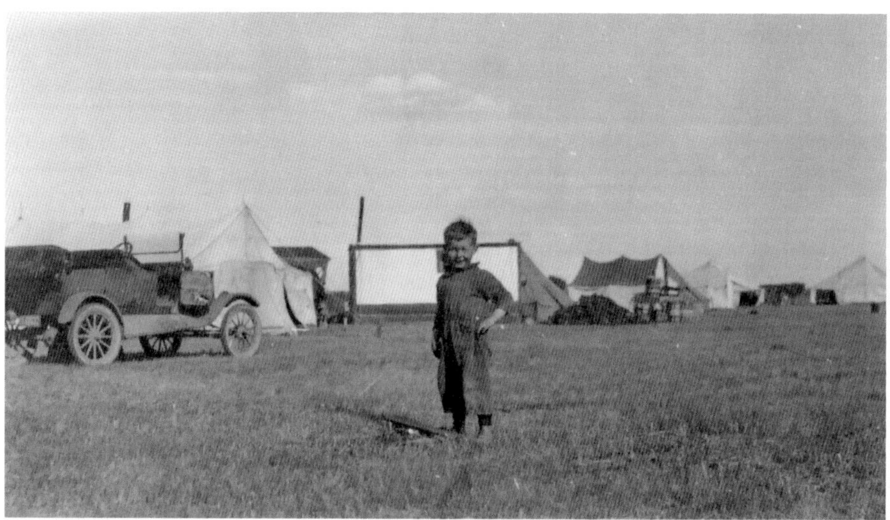

Families often accompanied road crews on construction projects, sometimes in remote areas of the state. *MDT.*

for use improving the highways for distribution among the state highway departments." The states had to pay the freight charges for shipping the equipment to them and agree not to resell it. The BPR allocated the surplus war equipment to the states in June 1919.

The Montana State Highway Commission was quick to take advantage of the government's offer and procured approximately $2 million worth of the equipment. Its imminent arrival prompted Conley to relocate the commission's shop from the state fairgrounds in Helena to the state prison's property near Deer Lodge. Conley announced to the counties and the contractors that the equipment was available to them for a nominal rental fee. Those wishing to rent equipment paid the shipping cost from Deer Lodge to its destination. Spare parts and repair work were available through the commission's Deer Lodge shop. Conley required those obtaining more than two trucks or other pieces of equipment to pay the wages of a highway department employee, who would oversee the fleet and work as a truck driver.

In all, the highway commission obtained 383 trucks. The commissioners distributed trucks to the counties, contractors and the convict road crews. Many of the trucks were Nash Quads equipped with mechanical hoists and dump bodies for use as gravel haulers. The Nash Quads were four-wheel-drive vehicles produced by the Jeffrey Motor Company and Nash Motors between 1913 and 1919. One Montana operator described them as "a huge hauling machine, a kind of petrol eating dinosaur." Many found the Quads difficult to drive with their four-wheel drive and four-wheel steering. The truck had a four-cylinder engine, a hand-cranked starter with no windshield, no lights and a top speed of eighteen miles per hour. The tires were solid rubber and "virtually indestructible, often outlasting the truck."[17]

Mineral County was the first to acquire the surplus trucks and eventually had eleven of the monsters in its fleet by the mid-1920s. It used the trucks to construct its portion of the Yellowstone Trail (later U.S. Highway 10). The rugged terrain of the route, coupled with mishandling by the drivers, required frequent repairs to the trucks. The county destroyed so many truck springs that Conley wrote the county commissioners a blistering letter about the problem:

> *We have sent you nine springs already. There is something wrong with your drivers or your roads. We have sixty-five trucks out and you are the only ones who have called on us for springs...Think if you have to buy springs you'll be more careful with them. We don't propose to furnish you with any more springs if you are going to use them like this.*[18]

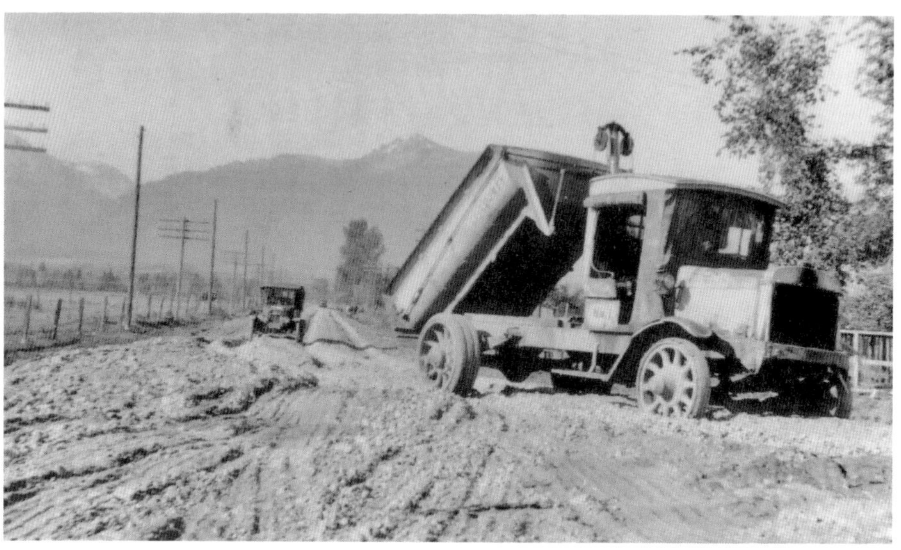

Described as a "petrol-eating dinosaur," war surplus Nash Quads were used on road construction in Montana after Word War I. *MDT.*

Along with the Nash Quads, the commission also acquired automobiles, ambulances, motorcycles, 117 wagons, seventeen twenty-ton caterpillars and an Austin Tandem Steamroller.

In April 1919, the highway commission, along with the commissioners from Carbon, Big Horn and Dawson Counties, awarded the first contracts for projects utilizing Federal Aid; the first let in Montana under the 1916 Federal Aid Road Act. They included the construction of nearly 3 miles of the Black and White Trail in Carbon County (the forerunner to the Beartooth Highway), a two-mile section of highway between Custer and Hardin and sixteen miles of the Red Trail in Dawson County. By the end of 1919, the highway commissioners and fifteen boards of county commissioners awarded twenty contracts worth over $900,000. The projects encompassed 98 miles of the 7,700-mile state highway system in Montana.

Once the highway and county commissioners awarded the contracts, however, construction work did not always go as planned. Highway construction in Montana was a new endeavor, and none of the Montana-based contractors had much experience building roads or owned the proper equipment to build them. Their inexperience resulted in "less than satisfactory progress as well as higher cost of supervision" for the highway department. Labor and material shortages prevalent during World War

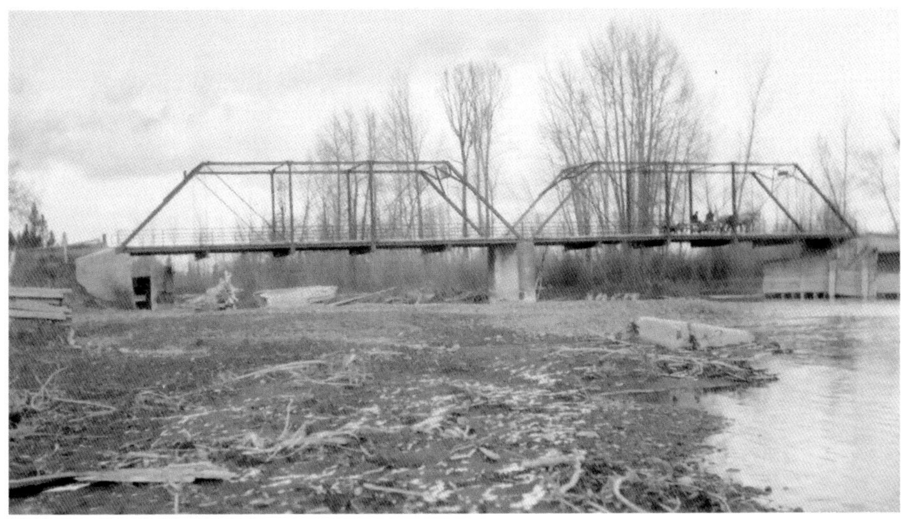

Built in the 1890s, the Whitefish River Bridge near Kalispell demonstrates that early steel bridges in Montana were intended for horse and wagon traffic. *MDT.*

I persisted into the 1920s. Although the commission enacted a policy that gave hiring preference to locals on highway projects, the population decline in eastern Montana after the war made labor scarce, and those men hired by contractors were often "grossly inefficient," hardly worthy of the relatively high wage rate of $6.50 per day common in the state. The highway commissioners complained of slow progress on the projects, with the counties either unwilling or unable to push the contractors to finish their projects by the deadlines specified in the contracts. Of the twenty highway and bridge projects contracted in 1919, none was completed by the end of the year; fifteen projects were still incomplete by late 1920.[19]

Concurrent with the road construction projects, the highway department surveyed 343 miles of roadway and prepared detailed plans for 46 miles of highway. During the summer of 1919, it also investigated a little over 3,992 miles of Federal Aid highways in Montana for future construction, concentrating on grade and alignment, drainage and availability of surfacing materials. The engineers adopted gravel surfacing as the "type best representing the traffic requirements of the state," and the highway commission would not accept contract bids for anything below that standard. Although gravel surfacing was the standard, the highway department developed standards for scoria-surfaced roads in eastern Montana, where good gravel sources were often scarce or too far away from the construction

Modernizing Montana's Bridges

While the Montana State Highway Commission had only minimal influence on the state's early road-building efforts, the same couldn't be said of its bridge department. One of the primary reasons the legislature created the highway commission in 1913 was to provide a level of control over bridge building in the state. Prior to the formation of the state's Bridge Bureau, county surveyors furnished the general layout of the types of bridges needed, but the bridge contracting companies supplied the designs. Many times, a county overpaid for a bridge or got something that didn't meet its needs. The counties were only nominally in charge of the construction process. The mish-mash of different bridge types, misrepresentation by some bridge companies and the bridge pooling issue directly influenced the Montana legislature's creation of a bridge department in the state highway commission in March 1915.

The state legislation required that the new Bridge Department standardize bridge designs, develop bridge contract packages for the counties and provide a method to ensure the counties got what they paid for by placing state bridge engineers on the projects. Within two months of the formation of the department, the highway commission hired Charles A. Kyle as the state's first bridge engineer. Born in Canada in 1864, Kyle immigrated to the United States with his family in 1869. A civil engineer, he gained practical experience in bridge design and construction while working for the Baltimore Bridge and American Bridge Companies in the early twentieth century. He was working as a bridge engineer in Salt Lake City when the Montana Highway Commission hired him in May 1915.

Kyle's background in bridge design and connections to the bridge companies enabled him to quickly develop designs for steel, reinforced concrete and timber bridges within just a few months of his employment. It is likely that Kyle's designs were developed in other states, and he adapted them for use in Montana. Importantly, Kyle dispensed with the pin-connected design popular in Montana since the early 1890s in favor of riveted steel Warren trusses. Originally developed for the heavy loads of the railroads in 1848, engineers had repurposed the design for automobile traffic by the second decade of the twentieth

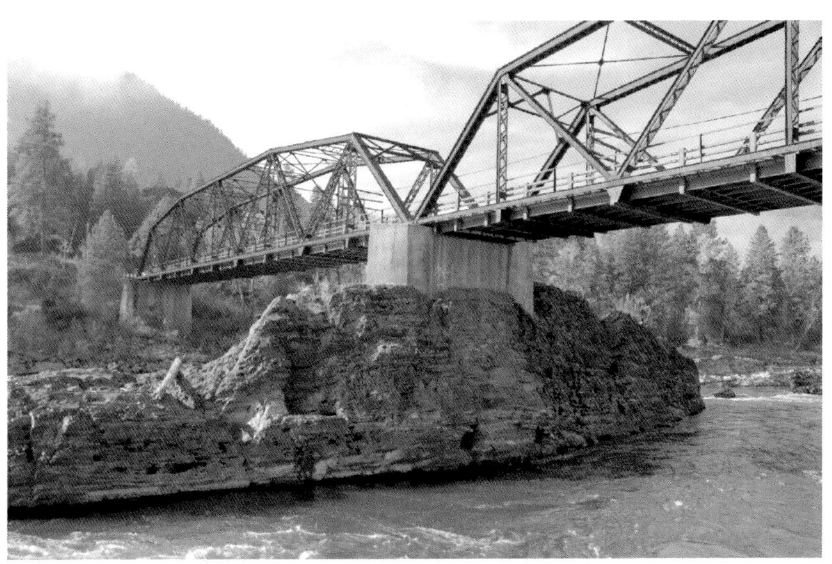

The Natural Pier Bridge across the Clark Fork River near Alberton. *Kristi Hager/MDT.*

century. The Warren truss was the standard truss bridge built on Montana's highways until 1946. Kyle also standardized bid packages for bridges and traveled throughout the state to consult with county commissioners about the new system. Unfortunately, the legislature didn't make mandatory county adherence to the bridge department's plans. While most counties followed the new system, a few did not.

The counties let the first state-designed and supervised bridge projects to contract in August 1915. Over the course of the next year, they awarded sixty-nine contracts for state-designed bridges. The first contract was for a single-span truss bridge across the Bitterroot River near Florence. Musselshell County let contracts for eight bridges across the Musselshell River the following month. Others were the Dickey Bridge in Silver Bow County, the Natural Pier Bridge in Mineral County, Browne's Bridge over the Big Hole River north of Dillon and contracts for small reinforced concrete bridges in Silver Bow and Fergus Counties. While many of these bridges no longer exist, Browne's Bridge still stands and is the earliest remaining state-designed bridge in the state.

Fred Burr and James Minesinger constructed a toll bridge at the Big Hole River crossing of the Bannack–Deer Lodge Road in 1863. Joseph Browne, a miner, bought the bridge from the men, and the first territorial legislature granted him a license to charge travelers for its use in 1865. Even though most of Montana's counties had assumed control of the state's toll facilities by 1892, Browne operated the toll bridge until his death in 1909. Beaverhead and Madison Counties accepted joint ownership of the bridge two years later. By 1915, old age, high water and a fire had forced both counties to condemn the old bridge. Residents continued to use it despite the center span's tendency to sway whenever any weight was placed on it. A new bridge was the counties' top priority under the 1915 legislation, and Kyle completed the plans for the structure that year.

In September 1915, the state highway commission and Beaverhead and Madison County commissioners awarded a contract for a new bridge to Missoula bridge builder O.E. Peppard. The *Dillon Examiner* reported that the new bridge would "be of sufficient strength to bear

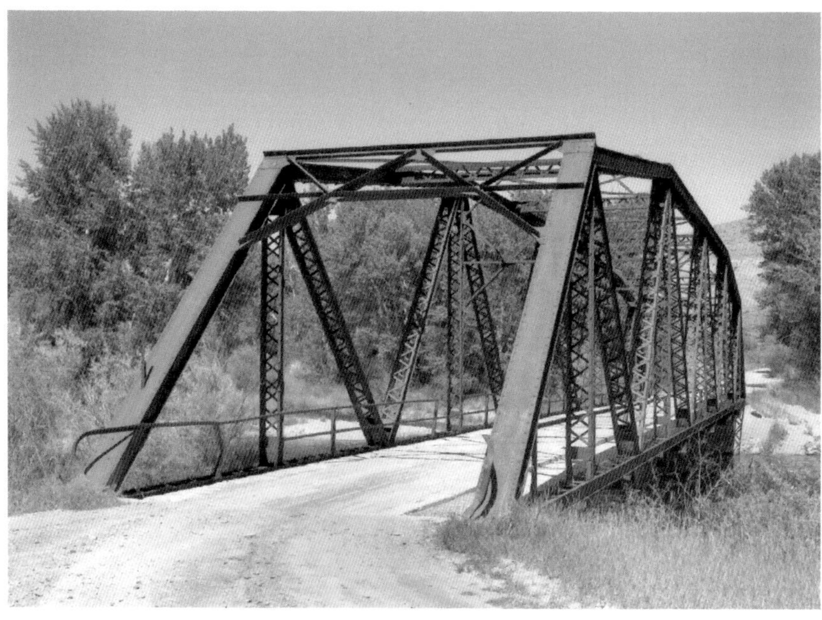

Browne's Bridge is one of the first State Highway Commission–designed structures built in Montana. *MDT.*

the weight of a 20-ton steam roller and will be one of the heaviest bridges in the county, containing about 140,000 pounds of steel." In February 1916, Peppard temporarily stopped construction of the bridge because he could not obtain the steel necessary for the superstructure. World War I caused a boom in steel production in the United States as mills supplied steel to Great Britain and France for the war, making it a scarce commodity for the domestic market. Peppard eventually obtained the necessary steel and resumed construction of the bridge, completing it in late March 1916. Highway commissioner George Metlen visited the bridge in March and pronounced the structure "one of the best bridges that has been built in the county and that it should last for an indefinite length of time." Within weeks of completion of the new bridge, high water washed out the old toll bridge, which had for many local residents become a "melancholy reminder of the passing of the old west, and its pioneer men and their works."[20]

site. The engineers also developed standards for macadam-surfaced roads. In 1919 and 1920, contractors built nearly 7 miles of hard-surfaced roads in the vicinity of Billings and Bozeman.

The 1916 Federal Aid Highway Act laid the groundwork for how federally funded highway construction projects would occur for the next three decades. The formative 1913 to 1920 years thrust the state into the leadership role for highway construction by establishing design standards, providing information about modern road practices to the counties and laying the groundwork for federal involvement in the road-building process. The transition was not without its problems; the state struggled with how to implement an efficient process and develop partnerships with the local and federal governments. The legislature created the highway commission as a purely advisory body in 1913 and had assumed a proactive role in the state's road-building program by the end of the decade. That role would intensify during the 1920s and ultimately result in the commission and highway department assuming full engineering and financial responsibility for developing the state's highway system.

Part III

THE MONTANA HIGHWAY DEPARTMENT TAKES THE REINS

If Montana is to become a dairying and diversified farming state, irrigation and highways must engage our serious attention. The only way to secure a state system of roads is through building by the state. The license tax and the gasoline tax money…should go into state road building and maintenance. Montana must get a policy in this regard or it can hardly expect to get anywhere.
—O.S. Warden to Governor Joseph M. Dixon, March 18, 1924

During its first decade, the Montana State Highway Commission and the Montana Highway Department established themselves as key players in the state's road-building efforts. The commission developed policies while the department surveyed and designed projects and oversaw the contract bidding process. Despite their roles in road building, the commission and the department still functioned mostly as overseers to the counties to ensure that roads and bridges met national design standards. The counties, not the State of Montana, provided the matching funds for federal money. That detail was an enormous obstacle in the development of an interstate highway system. The relationship between the state and the counties significantly changed during the 1920s. Federal regulations increasingly placed more responsibilities on the state to guarantee that federal dollars were efficiently spent on the construction and maintenance of a modern highway system. By the end of the '20s, new tax sources had enabled the highway commission to provide state-matching money for Federal Aid funds, which finally enabled the creation of interstate highways.

The '20s were anything but roaring in Montana. A severe economic depression hit the state in the wake of World War I. That, coupled with a devastating drought in eastern Montana, caused an exodus of over 200,000 people from the Treasure State. With the declining population and economy, the counties found themselves faced with diminishing tax revenues, which severely hampered their ability to provide matching funds for federal money to build roads. While federal funds remained steady, state funds for the highway department's overhead, engineering and administration costs came from the state license plate law. The state legislature distributed most of that revenue to the counties. Financing road construction depended entirely on the counties and their ability to raise money through bonds and taxes. Despite the declining population, the number of automobiles in Montana dramatically increased from 58,785 in 1921 to 136,997 in 1930. The increasing number of motorists put pressure on the counties and the state to provide better roads at a time of declining revenues.

Like the counties, the highway department grappled with significant problems in the early 1920s. Material and labor shortages plagued the construction industry in the years following the Great War. The costs of construction materials—namely steel—rose dramatically. The poor quality of the work by some of the contractors disappointed the highway commissioners. Chief highway engineer John Edy pushed contractors to complete their projects in a timely manner but reported that by the end of

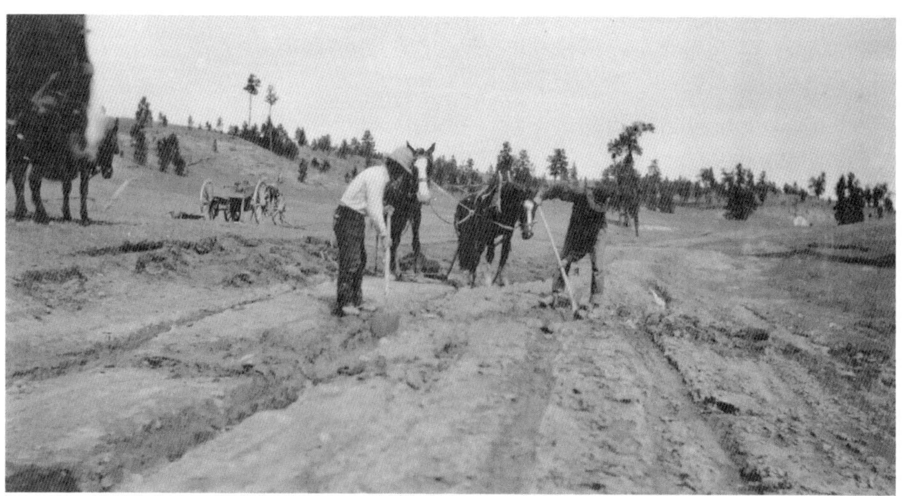

Road construction between Klein and Roundup in Musselshell County, 1920. *MDT.*

1920, of the "$5.8 million of contracts let; only $3 million worth of work was actually done."[21]

There were several reasons for the slowness of the work. The counties were responsible for purchasing highway right of way but didn't always obtain it before the highway commission let the contracts. Contractors often had to schedule their work around the fact that right of way had not yet been acquired. Importantly, some contractors didn't follow the specifications developed by the highway department's engineers. Between 1920 and 1930, there were 132 highway contractors active in Montana. Most were in-state companies, while only 18 percent consisted of out-of-state firms. One company, the State Highway Construction Company, was owned by former highway commissioners and highway department employees. While most of the contractors did not cause the department any trouble, there were a few exceptions. The contractor on the Skalkaho Road project accused the highway commissioners in 1923 of making changes to the road plans that extended the amount of time and money needed to complete the project. When the chief highway engineer disputed the claim, the contractor sued the highway commission. The dispute led to a protracted lawsuit and bad blood between the parties that continued for years after the commission settled the litigation in 1928.

With the highway commission's termination of the convict road projects in 1925 and the passage of the Good Roads Law in 1926, labor unions regularly appeared at the monthly commission meetings. In July 1927, representatives of the Montana Federation of Labor (MFL) sought assurances from the commissioners that they would enforce the Eight Hour Law and include a provision in future contracts guaranteeing that prevailing wage rates would be paid to labor in the counties where the work occurred. The MFL was the first labor organization to appear before the highway commission, but it would not be the last time labor union representatives petitioned the commissioners about labor issues. The commission was sympathetic to labor during the 1920s and 1930s. Some contractors habitually violated the Eight Hour Law in order to meet their contract deadlines, vindicating the unions' concerns.

The Montana Highway Department prioritized projects based on importance to the state's overall highway system and tried to improve the worst sections of highway first. Federal regulations specified that funds could be expended on the state highway system only until it enacted new legislation in 1921. County delegations at highway commission meetings were common during the decade. Because the counties controlled the purse strings, their requests carried a lot of weight with the highway commissioners, often at the expense of projects that should have been constructed first. Collaboration with the counties also had its own problems. All too frequently, a county asked the highway commission to initiate a Federal Aid project, "which, after considerable expense had been incurred by the department, the local officials later decide to postpone or abandon."[22] Sometimes counties were unable to finance their share of Federal Aid projects or did not cooperate with the highway department because projects were too costly. These problems made it difficult for the highway department to plan its road-building program in advance.

For the first half of the '20s, the highway commission was hampered by the lack of control over the construction program and the debate with the BPR on how best to solve Montana's road problems. The state depended on Federal Aid to construct roads, but it lacked the finances to build roads to national standards on its own. In 1928, chief highway engineer Ralph Rader summed up the obstacles the department faced:

> *Montana is the third largest state in the Union, Texas and California only have greater areas. Montana's population, however, is less than that of all except nine states, and is less than that of many of the cities of the United States. If you add to this the fact that a large part of the road mileage lies in the Rocky Mountains, where construction is very difficult and expensive, and also the fact that about thirteen percent of the total area of the state is non-taxable public land, you will have a fair picture of the general road problem of Montana.*[23]

Eventually, the problem would be solved by the enactment of the long sought-after gasoline tax that, coupled with new federal legislation, would give the commission the independence it desired. There would be other problems in the future, but the freedom from local control in road building marked a turning point in the highway department's history.

Highway construction was a political issue, as well as an engineering one. Governor Sam Stewart appointed Montana State Penitentiary warden Frank Conley to the highway commission in 1917. Conley had been the prison warden since 1890 and initiated an extensive convict labor program outside the prison walls. Prison labor had been building roads and bridges in Montana since 1910, and the highway commission advocated the use of prisoners beginning in 1913. Conley dominated the highway commission during his tenure. He formulated much of the highway commission's policy and determined where convict road crews would be used. In 1920, the commission obtained nearly $2 million in war surplus equipment, and Conley managed it heavy-handedly. He decided which counties and contractors leased it, often publically chastising those who abused the machinery. The surplus equipment issue likely contributed to Conley's removal from the highway commission by Governor Dixon in 1921.

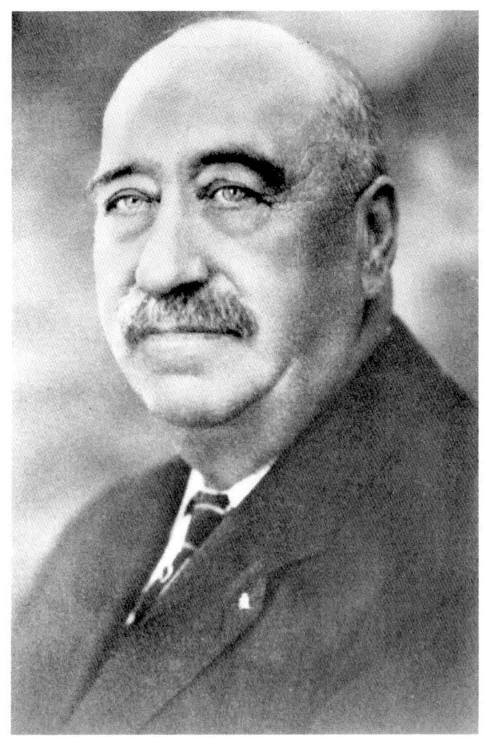

Frank Conley. *MHS Photograph Archives, Helena, 941-566.*

A Progressive, Governor Dixon sought to curb the power of the Anaconda Copper Mining Company in Montana during his term in office. He identified people in state government whom, he felt, had strong connections to the company; Conley was one of his first targets in 1921. After a lengthy and sometimes-ugly hearing, Dixon fired Conley, from both the highway commission and the penitentiary. Shortly after Conley's removal, his replacement on the commission, George Lanstrum, moved the highway department's shop back to the Helena fairgrounds. Conley, however, did not go quietly. In June 1921, he billed the commission for pasturage of its horses on his property in Powell County.

Governors Dixon and, later, John Erickson faced political opposition to their programs based on party lines in the state legislature. It was under Dixon's administration that the legislature enacted the first gasoline tax in 1921. The tax allotted a relatively small percentage of the tax revenue to the highway commission for administration, overhead and engineering costs, while the balance went to the counties to provide a portion of their matching for federal funds. Increasingly during the decade, however, legislators took tax revenue away from the commission and gave it to the counties. In 1925, Montana was the only state to return Federal Aid highway funds to Washington, D.C., because it couldn't raise enough money to match it.

The Federal Aid Road Act of 1921

The Federal Aid Road Act of 1921 appropriated $1.5 million for Montana on the condition that state funds were used to match federal funds. In an effort to increase the power of the state highway departments, it also specified that all construction and maintenance done on Federal Aid highways be performed under the highway departments' direct supervision. Most importantly, the legislation established the 7 Percent System, an effort to empower the highway departments by limiting the amount of mileage on which federal funds could be spent. The system consisted of 7 percent of the total mileage in each state. The BPR believed that the limited mileage would result in the rapid improvement of the interstate system. Federal funds expended under the 1921 road act concentrated on primary roads but also made provisions for secondary roads that connected county seats and federal roads within municipalities. In 1921, Montana could claim a total of 67,747 miles of county and state roads, of which 4,742 miles were on the state's 7 Percent System.

To administer the 1921 Road Act, the state legislature once again reorganized the Montana State Highway Commission. It dispensed with the cumbersome twelve-man commission in favor of a three-man panel. The legislature paid the chairman of the commission, George Lanstrum, a salary, and he functioned in much the same capacity as a business CEO

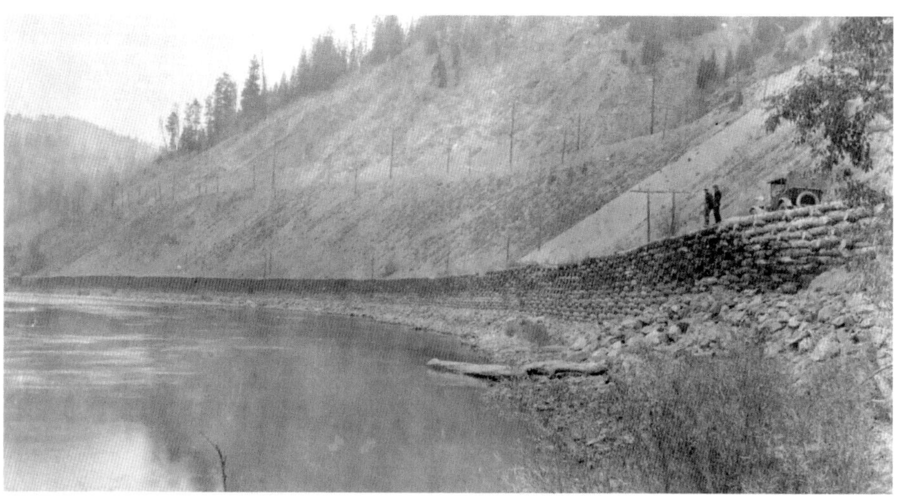

Engineers often designed highways using materials close at hand. Pictured is a log crib retaining wall supporting the Yellowstone Trail in Mineral County. *MDT*.

does today. The chief highway engineer and all department employees answered to Lanstrum. He determined policy matters, approved claims and "devoted all his time to the business of the office." The other two members of the commission held per diem positions and met with Lanstrum monthly to award contracts and establish policy.[24]

The highway commission expanded the highway department to accommodate the larger construction program under the new federal rules. The chief highway engineer continued to administer the department, assisted by administrators in the Bridge, Equipment and Testing Departments. In 1921, the department employed 164 people, including 50 employees in the districts. The main offices of the highway commission and department were in the state capitol building in Helena. The Bridge Department, however, was located elsewhere on the capitol grounds. The highway shop was at the Helena fairgrounds about four miles from the capitol. Limited space at the capitol meant constant jockeying among state departments to get premium spots in the building. In 1925, for example, Maggie Smith Hathaway of the State Bureau of Child Protection unsuccessfully attempted to acquire the rooms occupied by the highway commission.

Despite the expanded program, the department was vulnerable to the moods of the state legislature. Layoffs were commonplace during much of the decade if state revenue did not meet anticipated levels. Indeed, in 1929, the highway commissioners had difficulty even making payroll. Much of the lobbying done by the highway commissioners to the legislature involved creating a solid economic foundation for the operation of the highway department and the construction of highways under the provisions of the Federal Aid Road Act. In 1922, the commissioners proposed utilizing a gas tax, increased vehicle/license fees and royalties from federally owned oil lands within the state to institute an annual budget of $1.75 million. Out of that money, the department would pay administrative and maintenance costs. The money derived from the tax would be used to match federal funds, thus eliminating the counties from the process. Unfortunately, the commissioners' ambitious plan failed in the 1923 legislative session.

In March 1925, the American Association of State Highway Officials (AASHO) presented a resolution to the U.S. Department of Agriculture to develop a national highway numbering system. The plan included highways already on the Federal Aid system. To expedite the process, the committee held no public hearings but instead met with state highway departments

When Montana's Highways Had Names: The Trail Associations

A justification for the federal government's 7 Percent System involved the myriad trail associations then active in the United States. In 1921, there were over 250 trail associations in the country; 13 of those associations were active in Montana. In those days, roads had names rather than numerical designations. Formed mostly in the 1910s, the trail associations consisted of groups of interested businessmen, local boosters and other individuals who promoted their interests in terms of tourism and good roads. They designated linkages of county roads that connected national parks and other attractions that might draw tourists to a specific area. The roads passed through communities that advertised in trail association promotional material.

The associations gave colorful and evocative names to their roads, including the Yellowstone Trail, Theodore Roosevelt International Highway, the Y-G Bee Line, the Electric Highway and the Custer Battlefield Highway. They blazed the routes with brightly colored symbols. The Yellowstone Trail, for example, was marked by bright yellow circles with arrows pointing the way for motorists. The Vigilante Trail used the dreaded 3-7-77 numbers to mark its route in southwestern Montana. The trail associations received a boost in 1915 when the U.S. Army first allowed automobiles inside Yellowstone National Park.

Many of the roads designated by the trail associations were technically on the State Highway System designated by the state highway commission. Many of the associations published brochures or pamphlets to advertise the benefits of the route, provide reports on road conditions and advertise businesses along the way. They also lobbied the state highway commission to have federal funds spent on the improvement of the roads they promoted. Importantly, the trail associations also encouraged up-to-date engineering standards for their roads.

As automobiles became more sophisticated and road systems somewhat improved, more and more Americans took to the roads to tour the United States. Part of the stimulus for the growth of tourism in the American West was World War I. In 1917, the *Mineral County Press* wrote:

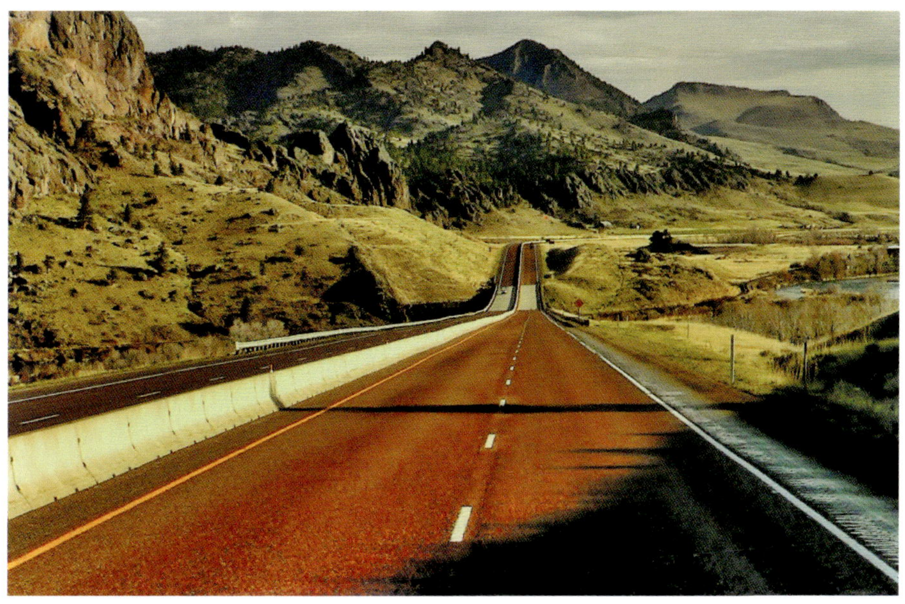

Northbound on I-15, about to cross the Missouri River just south of Great Falls. *Courtesy of Darrell J. McIntosh.*

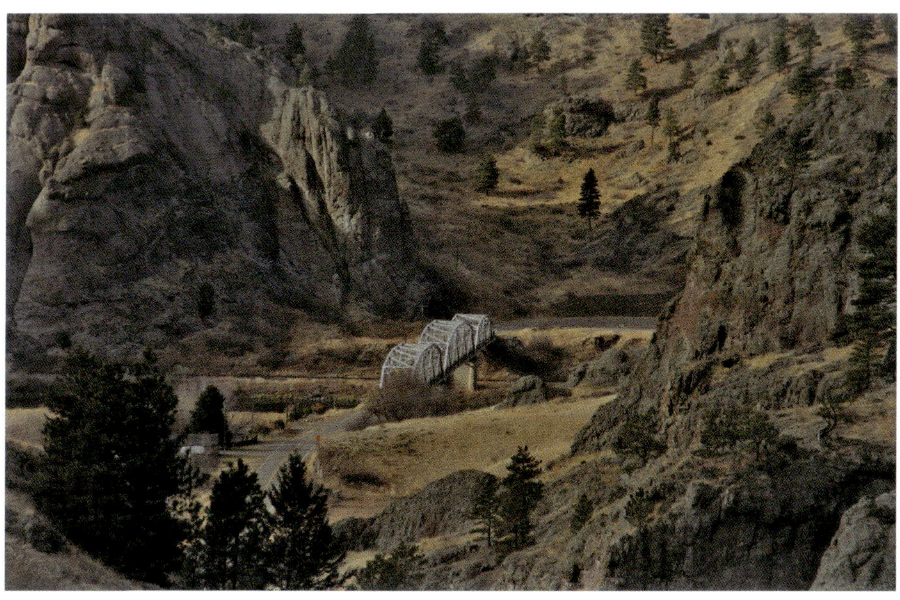

View from I-15 near Cascade. Bridge over the Missouri River. *Courtesy of Darrell J. McIntosh.*

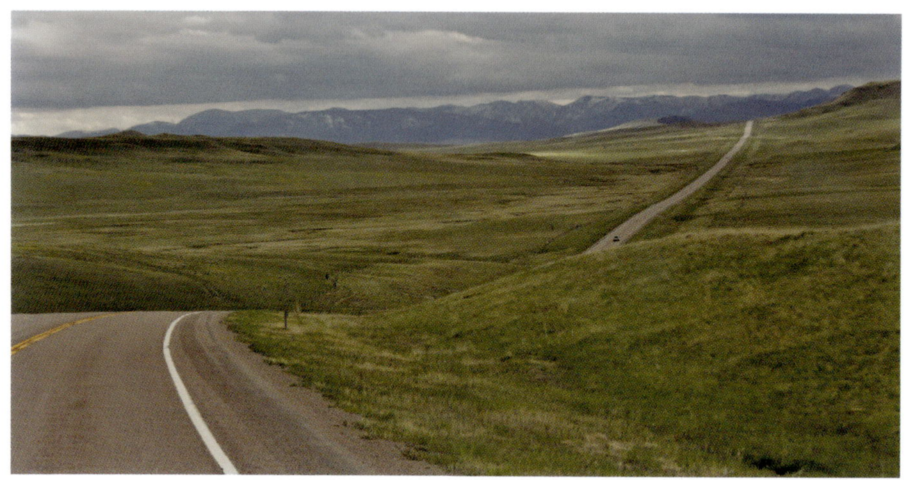

Southbound on Highway 200 near Great Falls. *Courtesy of Darrell J. McIntosh.*

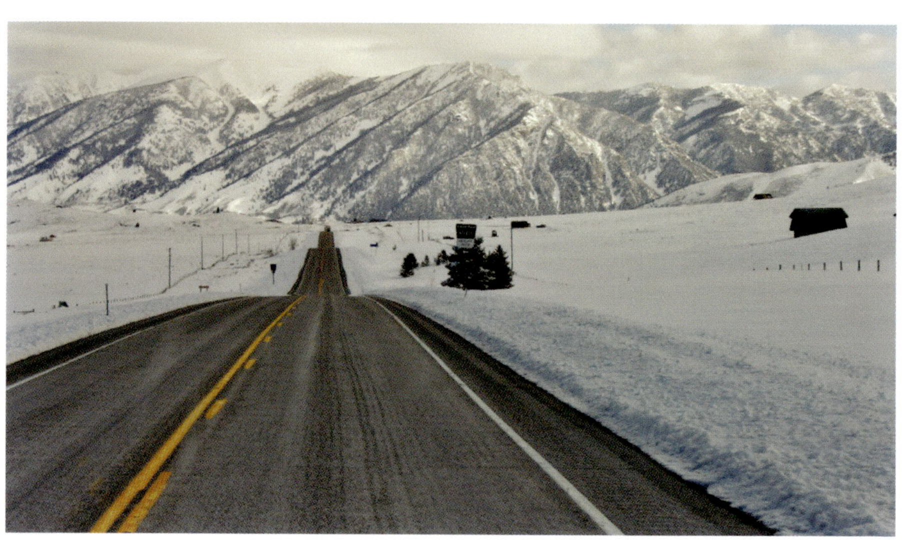

Just into Montana from Idaho on Highway 87, nearing northbound 287. *Courtesy of Darrell J. McIntosh.*

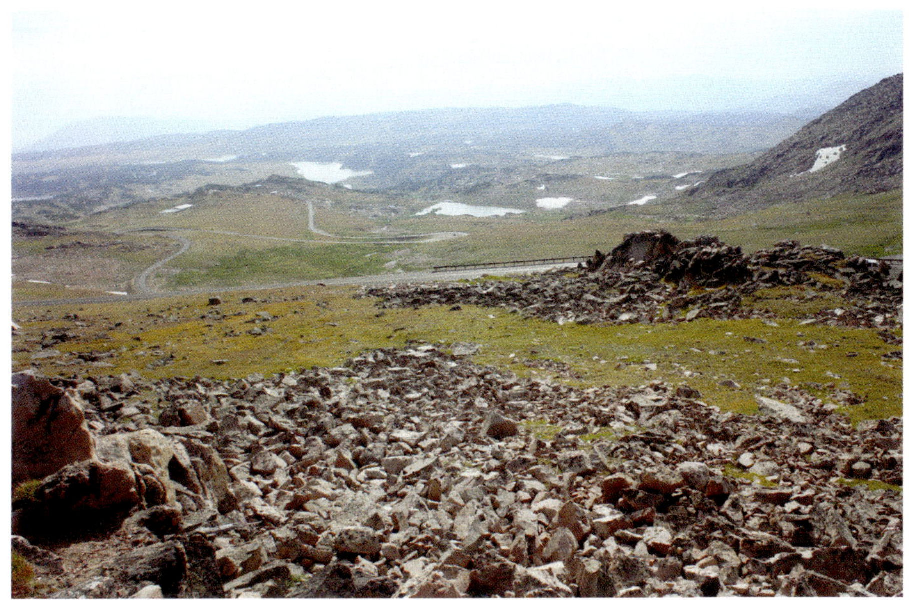

This page: Beartooth Highway. *Courtesy of Alan Davenport.*

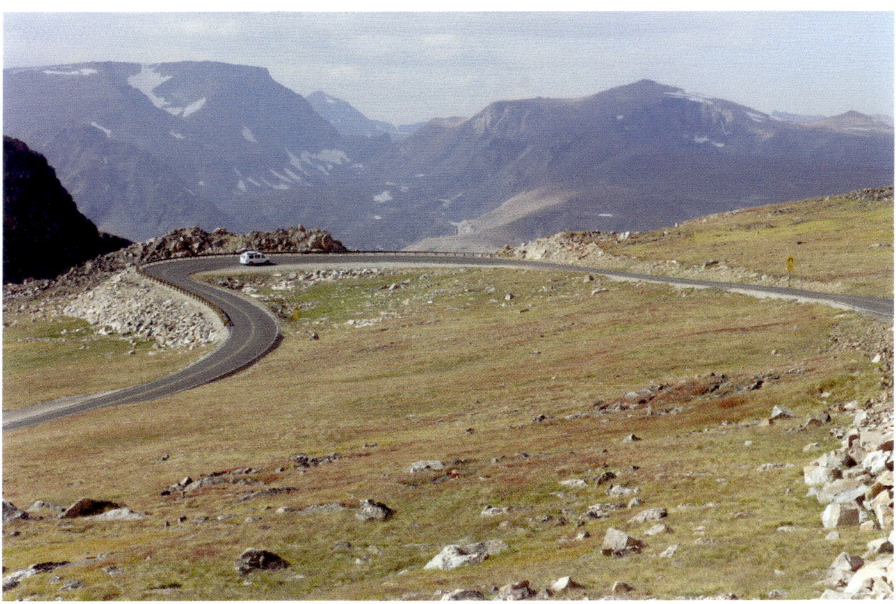

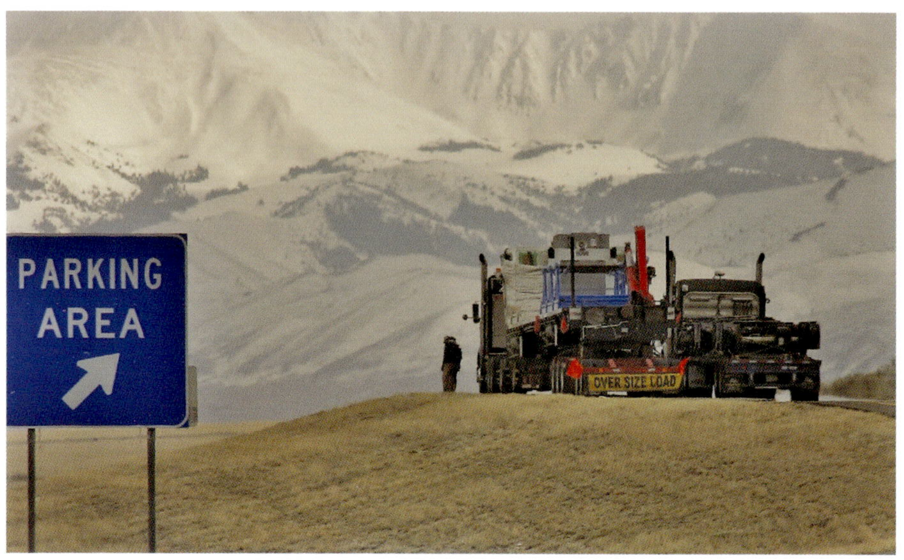

Southbound I-15 rest stop, just before crossing the Continental Divide into Idaho. *Courtesy of Darrell J. McIntosh.*

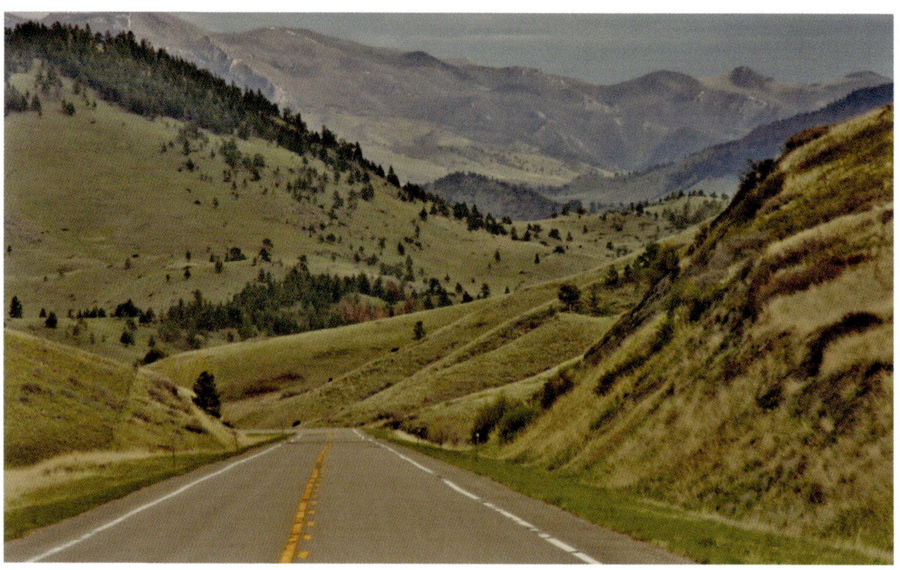

Above and opposite, top: Southbound on the very scenic Highway 287, nearing I-15 at Wolf Creek. *Courtesy of Darrell J. McIntosh.*

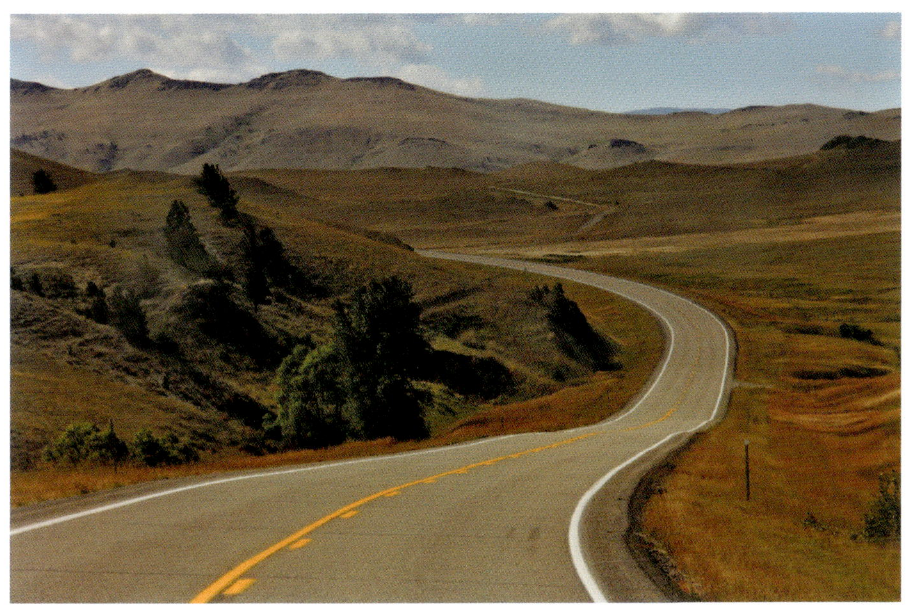

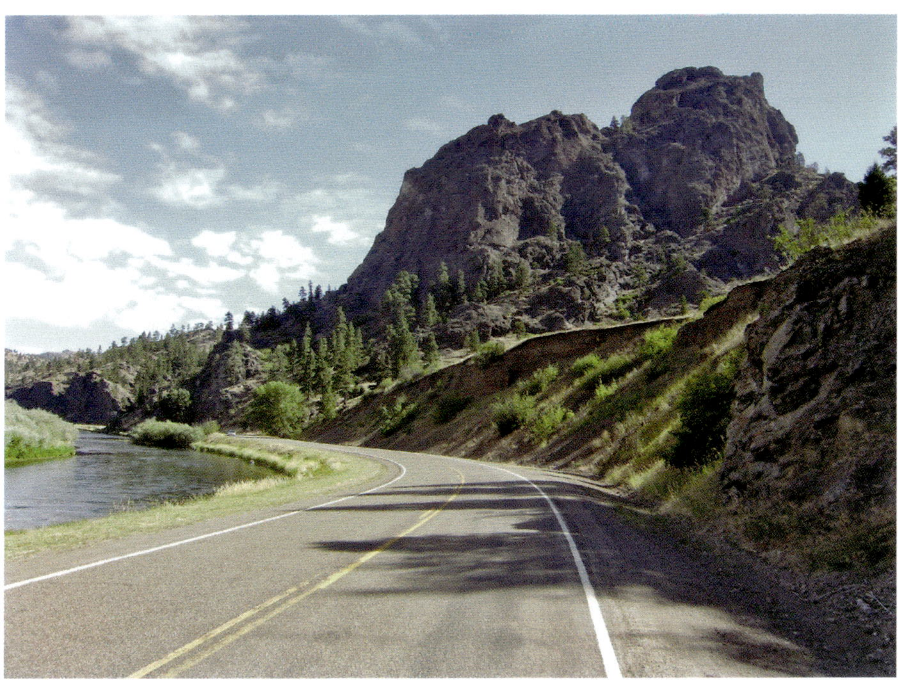

Old U.S. Highway 91 in Cascade County. *MDT.*

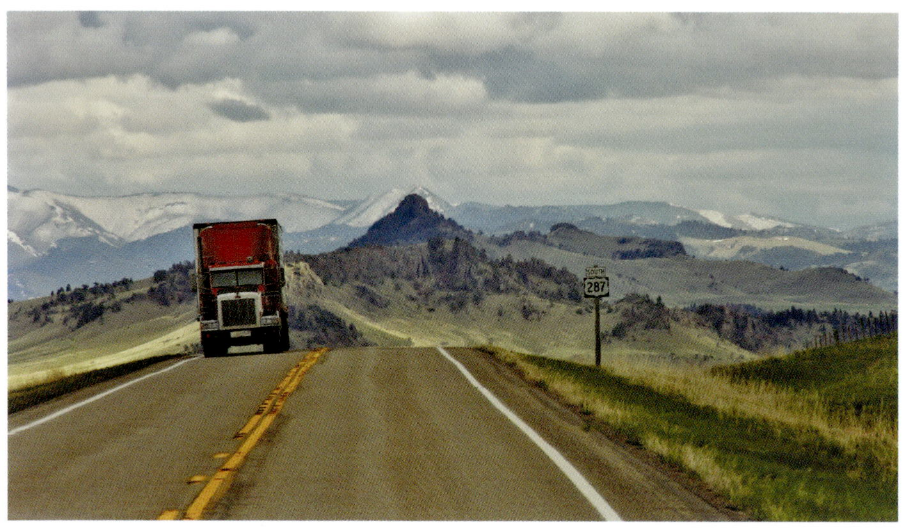

The view is looking southbound on Highway 287 nearing I-15 at Wolf Creek. *Courtesy of Darrell J. McIntosh.*

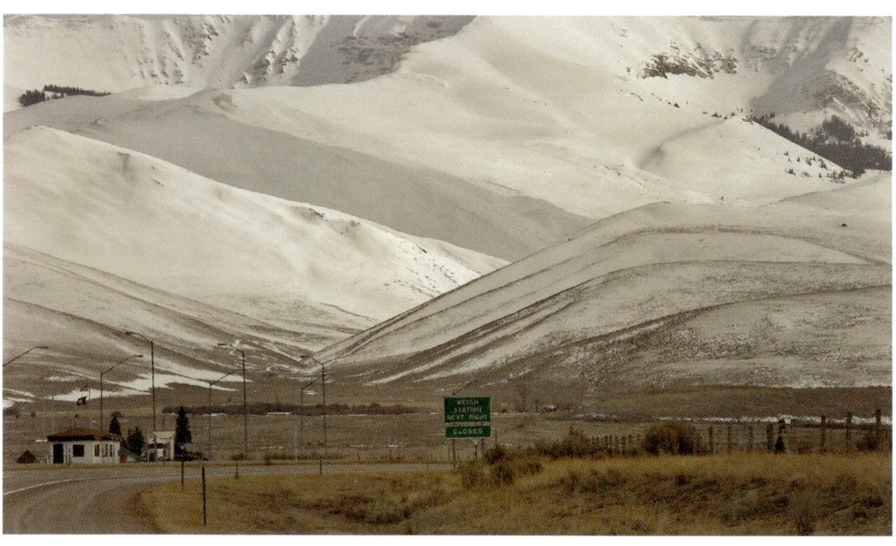

Southbound on I-15 at Lima, about to climb up and over the Continental Divide into Idaho. *Courtesy of Darrell J. McIntosh.*

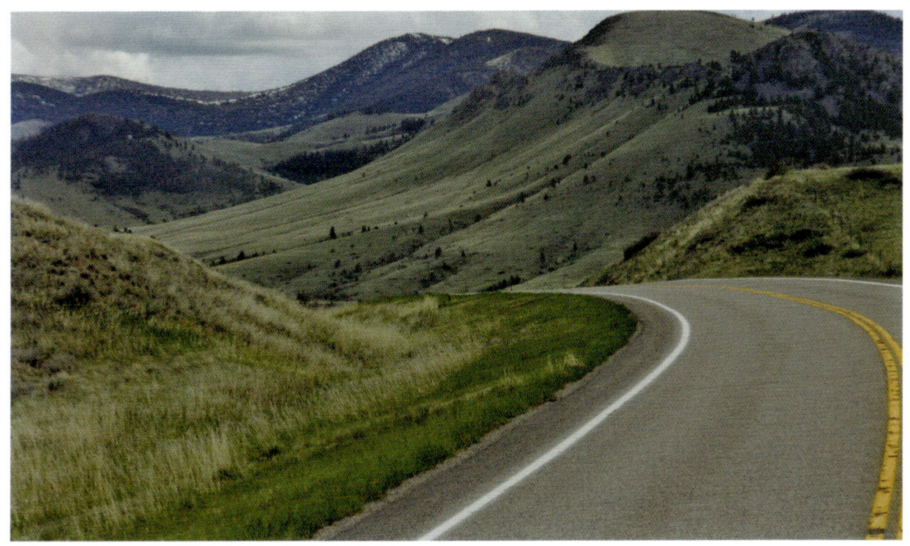

Southbound on Highway 287 nearing I-15 at Wolf Creek. *Courtesy of Darrell J. McIntosh.*

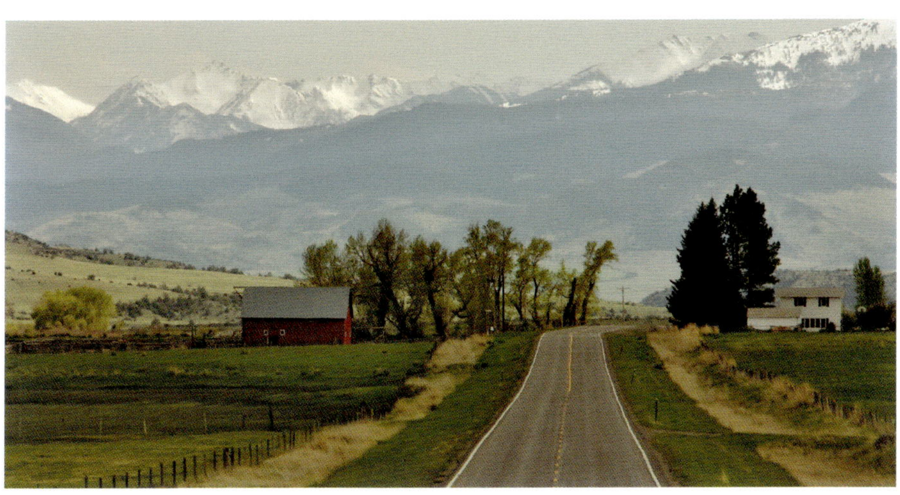

Southbound on Highway 41, south of Whitehall. *Courtesy of Darrell J. McIntosh.*

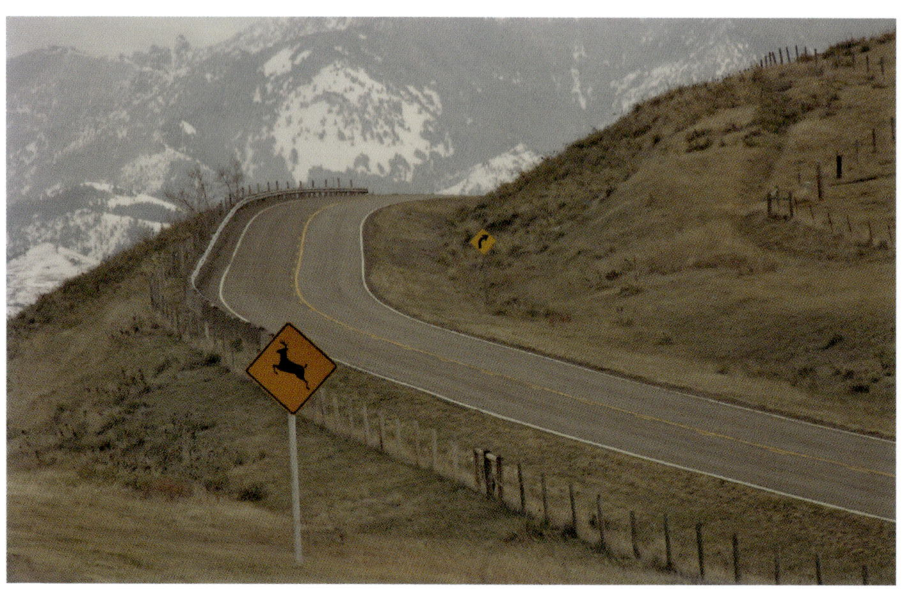

Southbound on I-15 near Cascade, showing a small side road. *Courtesy of Darrell J. McIntosh.*

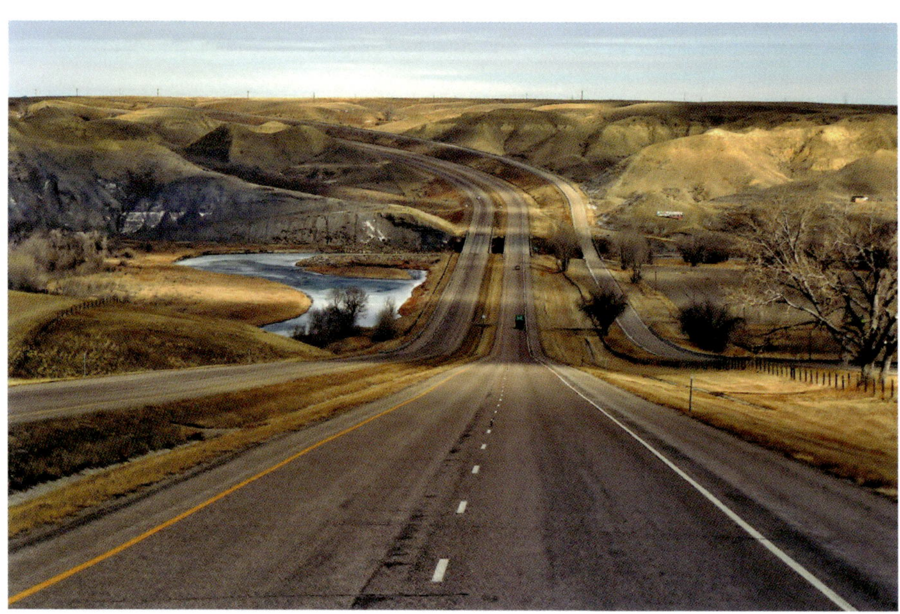

North of Great Falls on I-15, nearing the Marias River, just south of Shelby. *Courtesy of Darrell J. McIntosh.*

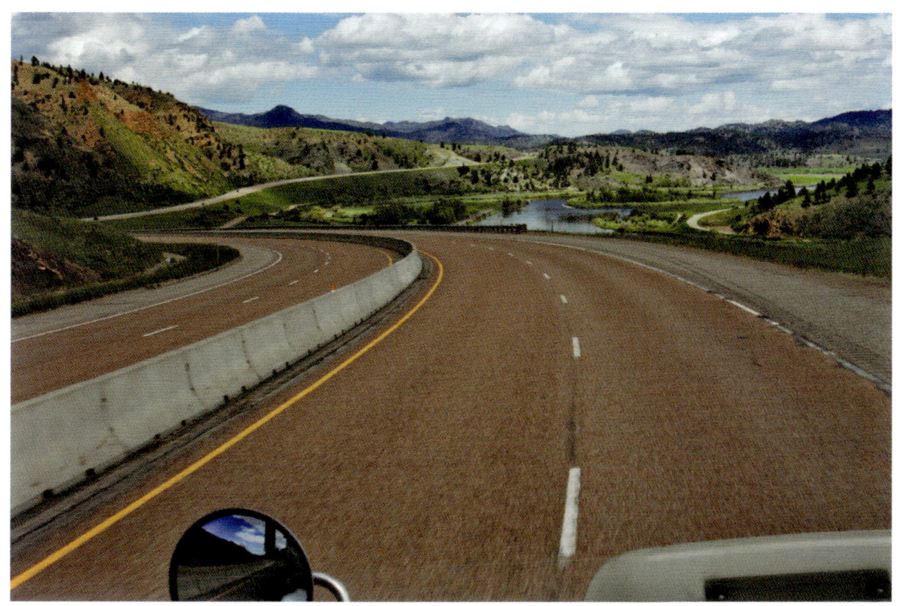

Northbound on I-15 heading toward Great Falls. *Courtesy of Darrell J. McIntosh.*

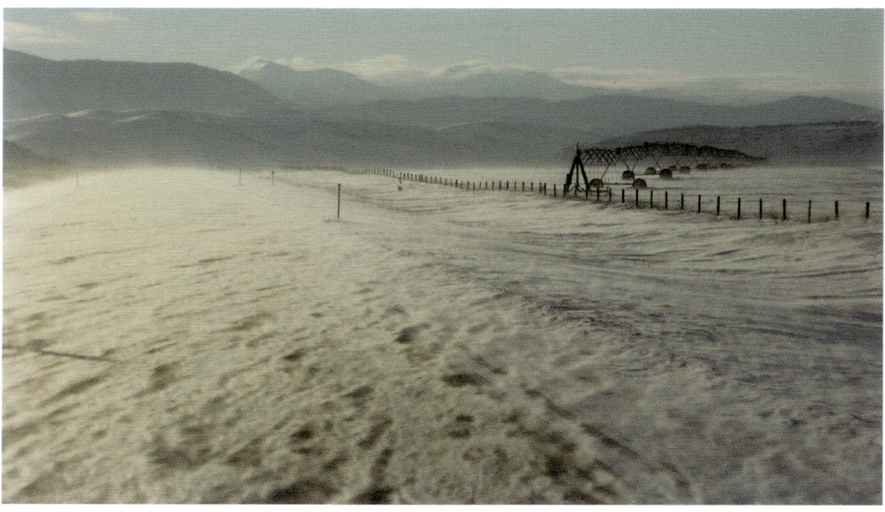

Southbound on Highway 287, south of Harrison, heading to Ennis. *Courtesy of Darrell J. McIntosh.*

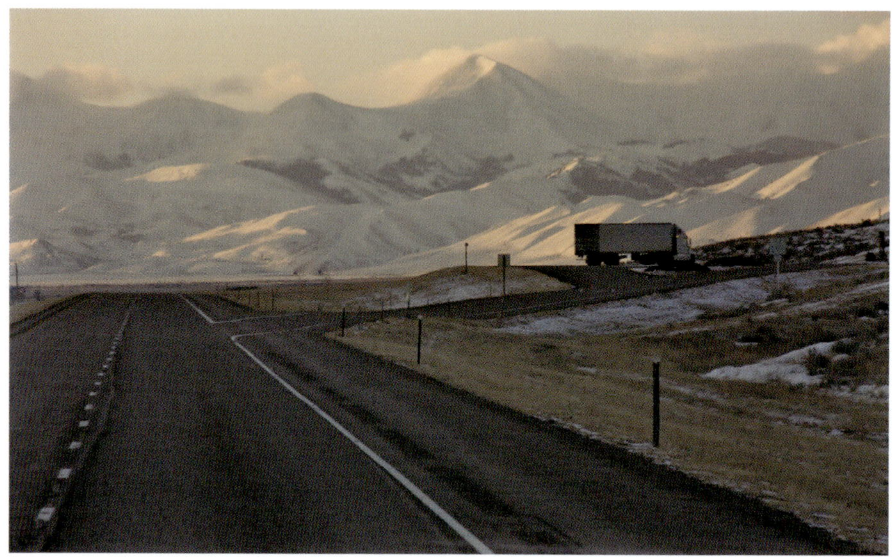

The southbound rest stop on I-15, south of Dillon, nearing the Continental Divide and Idaho. *Courtesy of Darrell J. McIntosh.*

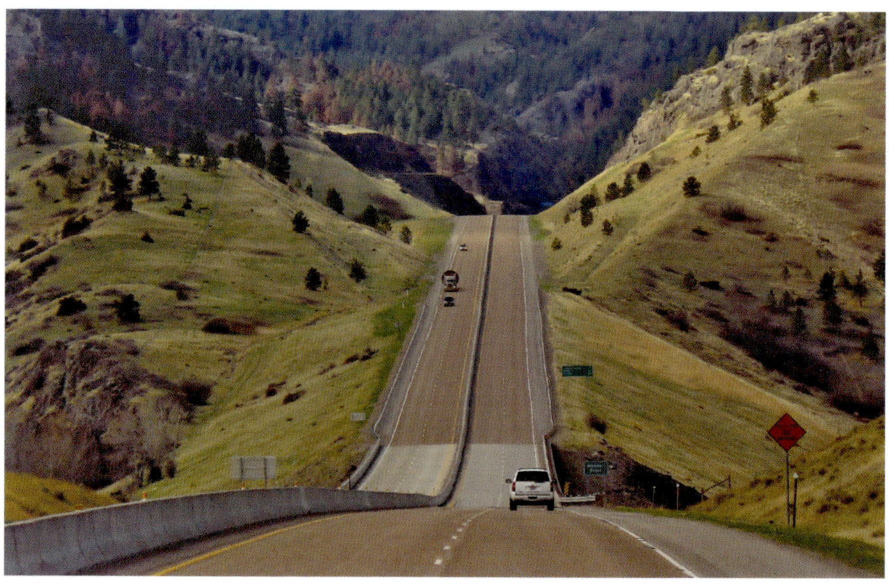

Just south of Great Falls, heading south on I-15, about to cross the Missouri River. *Courtesy of Darrell J. McIntosh.*

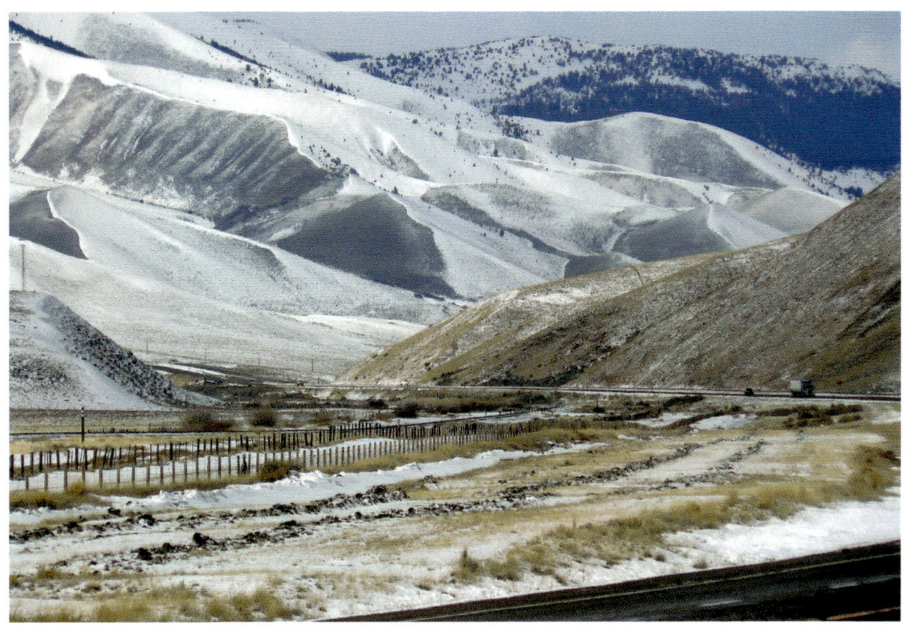

Just coming into Montana from Idaho on I-15, following along the Bitterroot Mountain Range. *Courtesy of Darrell J. McIntosh.*

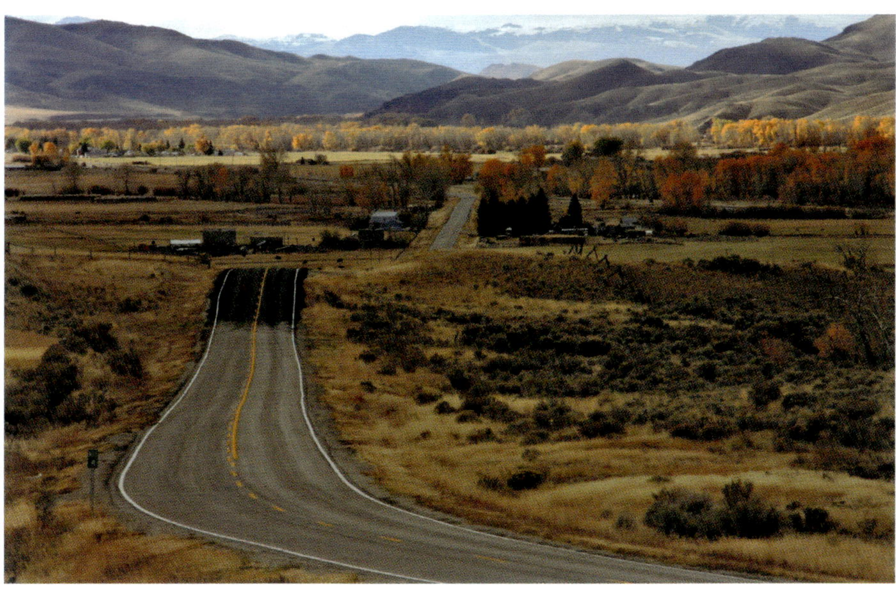

View from southbound on I-15, south of Butte, showing Highway 43 heading into Melrose. *Courtesy of Darrell J. McIntosh.*

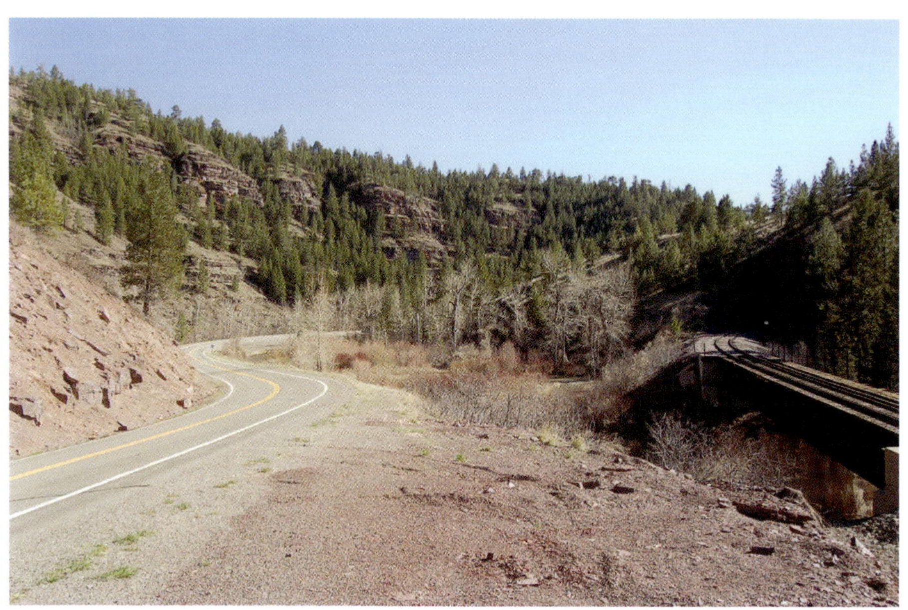

Built in the 1930s, old U.S. 91 through the scenic Prickly Pear Canyon follows the route of a wagon road established in the 1860s. *Courtesy of Jon Axline.*

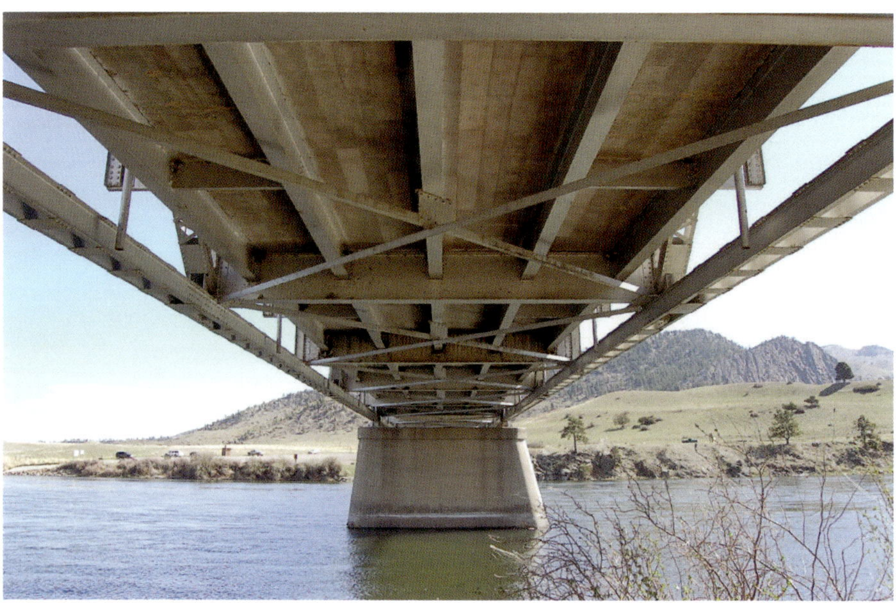

The underside of the Missouri River Bridge, northeast of Wolf Creek. *MDT.*

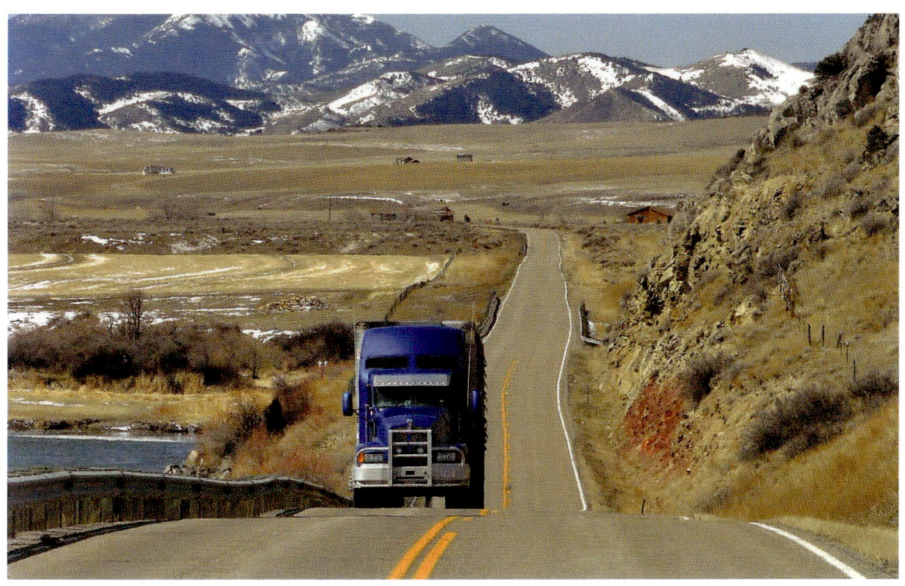

Heading north of Dillon on Highway 41 toward Twin Bridges and Whitehall. *Courtesy of Darrell J. McIntosh.*

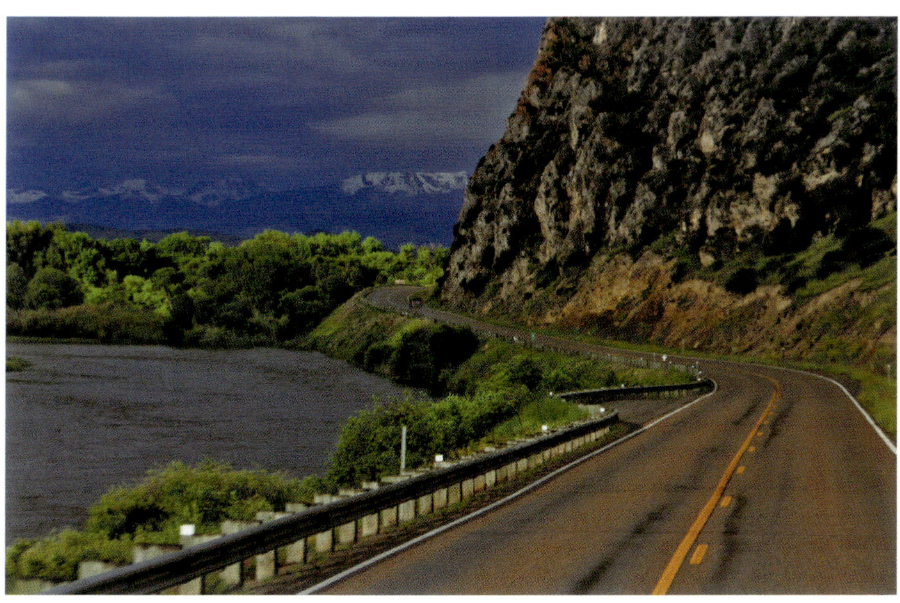

Southbound on the 287, following along the Jefferson River. *Courtesy of Darrell J. McIntosh.*

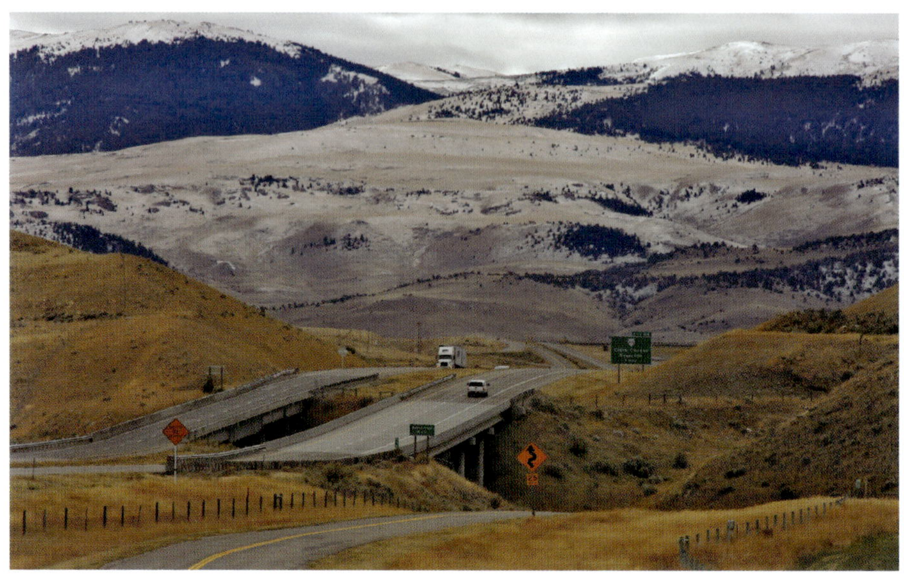

Southbound on I-15, south of Dillon, nearing the Clark Canyon Reservoir and Idaho. *Courtesy of Darrell J. McIntosh.*

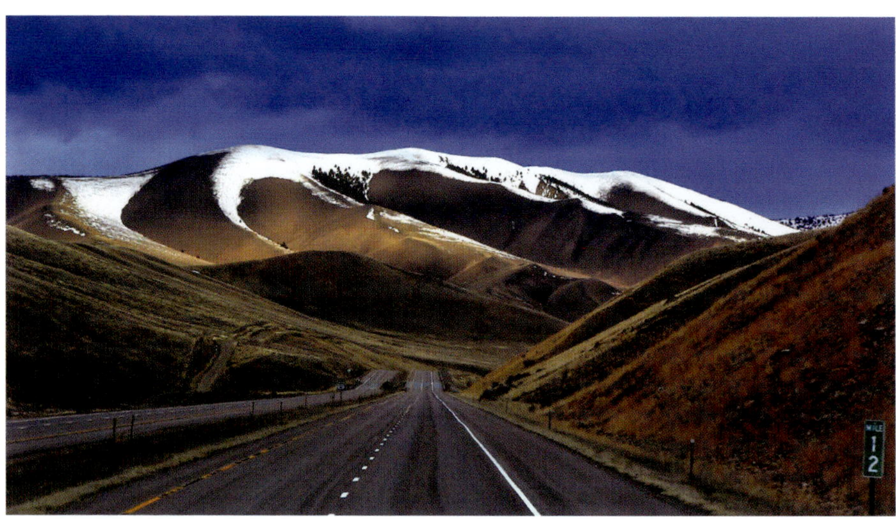

Northbound on I-15, nearing Lima and the Truck Scales. *Courtesy of Darrell J. McIntosh.*

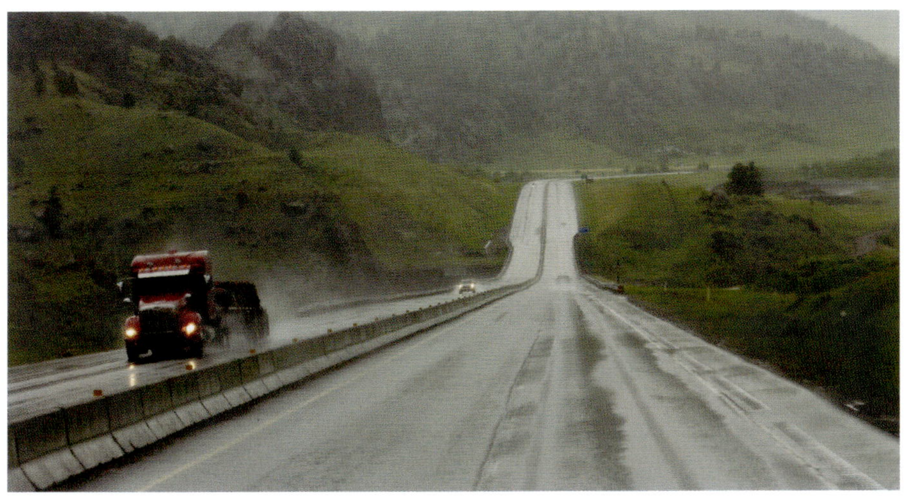

Northbound along the Missouri River on I-15, near Hardy Creek. *Courtesy of Darrell J. McIntosh.*

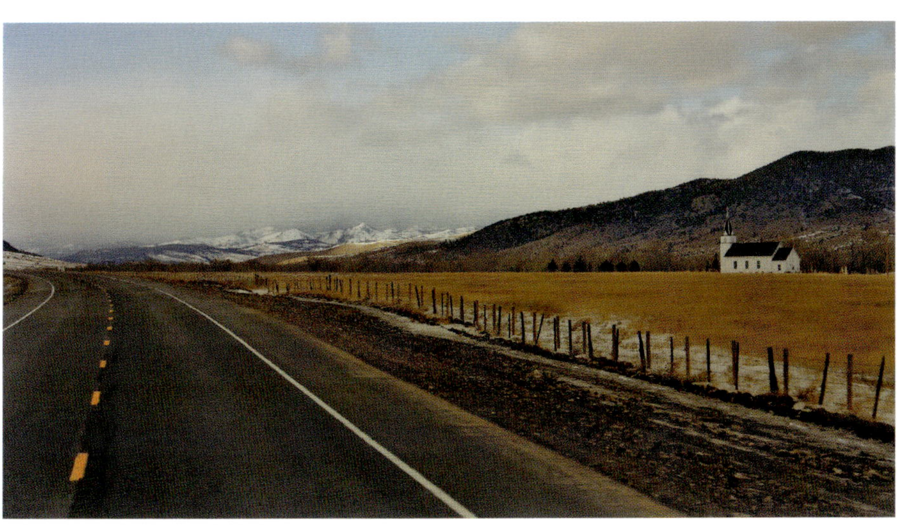

Back country on Highway 69, just south of Boulder and north of Whitehall. *Courtesy of Darrell J. McIntosh.*

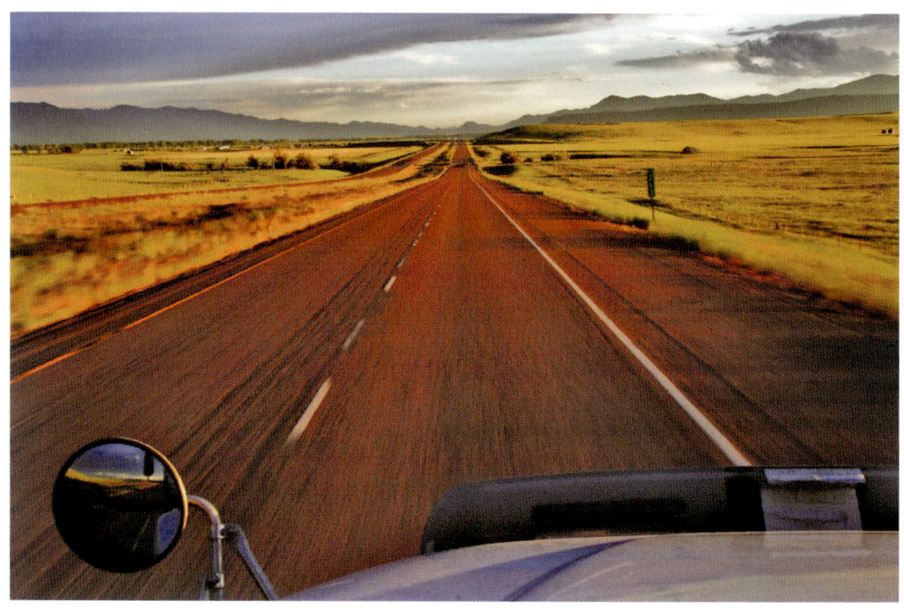

Southbound on I-15, just south of Great Falls. *Courtesy of Darrell J. McIntosh.*

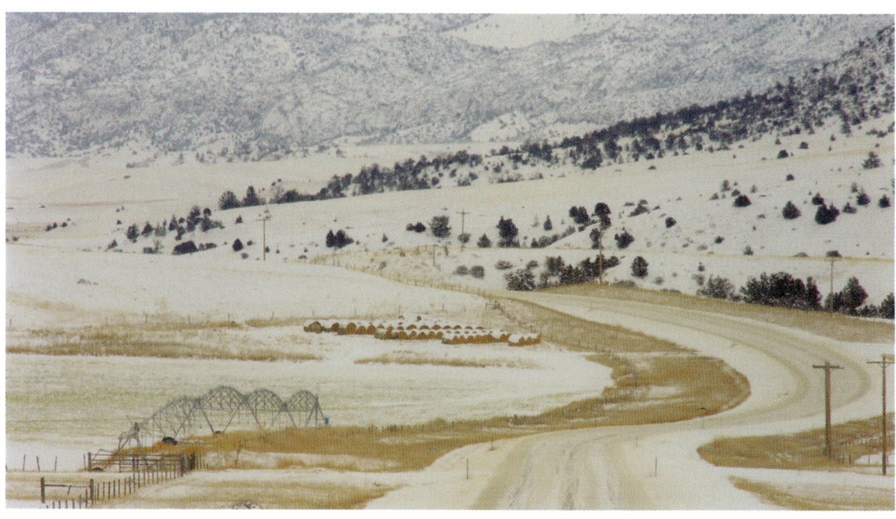

Northbound on Highway 69, near Boulder and Whitehall. *Courtesy of Darrell J. McIntosh.*

> *It is claimed that the people of Montana do not appreciate the value of tourist travel. Europe is the playground of the wealthy American. The war has shut him out of his playground. He is indulging his taste for travel in long automobile trips through the West, and automobile travel means the distribution of money all along the route of automobile travel.*[25]

Many promoters and businessmen across Montana and the United States shared the newspaper publisher's sentiments, seeing automobile tourism in terms of dollar signs.

The highway commission treated the trail associations as allies in improving the state's highway system. Much of Commissioner George Metlen's early efforts to develop the state's highways was directed at those groups. He regularly spoke to their conventions about the highway commission's efforts to improve roads and discussed other road issues with them. Both the highway commission and trail associations didn't have money of their own to actually build roads. They worked through local governments and businessmen to designate the routes, obtain funding for road improvement projects and generate public support for their efforts.

One of the most active trail organizations in Montana was the Yellowstone Trail Association (YTA). Formed in Ipswich, South Dakota, in 1912, the trail was the first interstate highway in the United States. It comprised an interconnected network of county roads that provided a connection between Minneapolis–St. Paul, Minnesota, and the north entrance to Yellowstone National Park. The YTA orchestrated a national media campaign promoting highway issues, lobbied state legislatures and local governments and actively pressed the U.S. Army to allow automobiles into Yellowstone National Park. At its first annual convention in Miles City, the association designed the Yellowstone Trail as a "good road from Plymouth Rock to Puget Sound." Local governments and businesses that were dues-paying members of the YTA maintained the roads. In return, the association advertised businesses and services along the route in its promotional publications. The YTA provided the model for other trail associations in the state. The Yellowstone Trail was the forerunner of U.S. Highway 10 and Interstates 90 and 94.

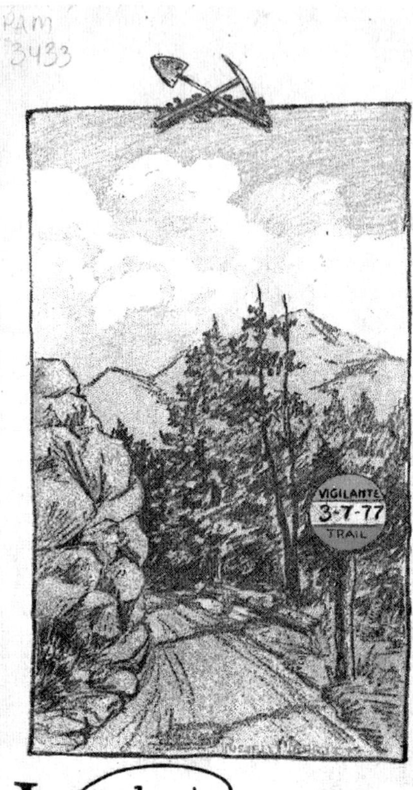

The Vigilante Trail was one of several named "tourist" highways that crisscrossed Montana in the 1910s and 1920s. *MHS Research Center, Helena.*

Within a decade of the creation of the YTA, at least twelve other "tourist highways" crisscrossed Montana. All were promoted by their own associations and devised distinctive symbols to blaze the way. The names made it obvious to motorists where the road would take them: the Park-to-Park Highway connected Glacier and Yellowstone National Parks, while the Custer Battlefield Highway provided a route to the American West's most famous battlefield. For others, it was a little more problematic. The Great White Way was intended to stir up images of the 1893 Chicago World's Fair, while the Electric Highway paralleled much of the route of the Milwaukee Road Railroad's electrified line in the Musselshell Valley. The associations published brochures or pamphlets to provide reports on road conditions and advertise businesses along the way. They lobbied hard the state highway commission to have federal funds spent on the improvement of the roads they promoted. Importantly, the trail associations encouraged up-to-date engineering standards for their roads. Competition for federal funds between the public and private entities was intense, with much of it playing out at the monthly highway commission meetings. The competition for improvements to the trails sometimes made programming by the highway commission difficult.

The trails began to fade away in 1926, when the Federal Aid highways received numerical designations. The Yellowstone Trail was redesignated U.S. Highway 10, while the Roosevelt Highway became U.S. 2 and portions of the old Blackfeet Highway became U.S. 89. For years, however, the Montana State Highway Commission continued to refer to some highways by their old names. It still made references to the Yellowstone Trail, Electric Highway, Y-G Bee Line and the Vigilante Trail in its monthly meetings until well into the late 1940s, when those names finally disappeared from its records.

at regional meetings. Initially, the committee identified 81,000 miles of highways that it recommended for numerical marking. It later pared the number down to 50,100 miles but, after meeting with the state highway departments, increased the number to 75,800 miles. In Montana, the mileage proposed for marking consisted of nearly 5,000 miles, including the Yellowstone Trail (U.S. 10), Roosevelt Highway (U.S. 2), Y-G Bee Line (U.S. 89), Custer Battlefield Highway (U.S. 87) and the Federal Aid system roads connecting the Idaho border at Monida to the Canadian border at Sweetgrass (U.S. 91). The federal government approved the numbering system in November 1926.

The highway commission, however, was slow to authorize the system. There was considerable debate at commission meetings about where to place signs, their design and who would manufacture them. Initially, the commission intended on marking only U.S. Highway 10. It considered contracting with the state penitentiary in Deer Lodge to manufacture the signs but soon dropped that plan and established a sign shop in Helena. Although there was a national standardized sign design, it wasn't mandatory that the states adopt it; the highway commission was unsure whether to use it or design its own sign. Eventually, it grudgingly adopted the national design. The department began installation of the route signs in 1929 and completed the project in 1930.

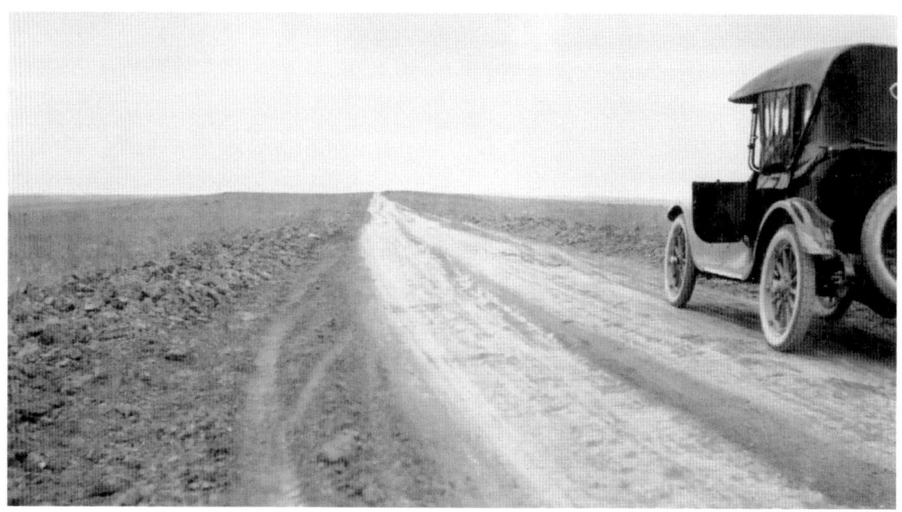

Montana's roads were hell on automobile suspensions and drivers' nerves as this segment of the Custer Battlefield Highway in Fergus County indicates. *MDT.*

Politically and economically, the first half of the '20s was a time of turmoil for the highway commission as it attempted to establish a working and efficient system of highway construction. In 1916, highway commissioner George Metlen recognized the problem of planning, financing and building roads as long as the counties controlled the purse strings. The Federal Aid Road Acts of 1916 and 1921 tried to remedy that situation by placing more control of the process with the state highway commission. But in Montana, that control would never be realized until the highway department became financially able to design and build highways. In 1924, new commission member Oliver S. Warden reported, "In the last two years, the counties had $3 million of Montana road funds and not less than twenty percent of those funds had been used to match federal funds." The state legislators, moreover, tended to side with the counties when it came to Federal Aid funds. They emphasized local control over state control and attempted to freeze the highway commission out of the process.[26]

The Federal Aid Road Act of 1921 sparked renewed efforts by the state to raise money to provide matching funds for federal dollars. Governor John Erickson, an ardent good roads supporter, told the legislature in 1925 that Montana needed $3.5 million to match accrued Federal Aid funds. Only by utilizing all of the gasoline tax and part of the license tax, Erickson stated, could the state raise that money. The legislature responded to his request by cutting the state highway fund's share of the money from 20 percent of the gas tax revenue to 15 percent. Legislative actions in the early 1920s virtually emasculated the highway commission and the highway department, leaving them only enough administrative money to maintain a skeleton organization. In 1924, just prior to his resignation from the commission, Chairman George Lanstrum declared, "If provision were not made for a state highway fund with which to match Federal Aid in the next two years, there would be little else for the State Highway Commission to do but wind up present road contracts and go out of business."[27] Importantly, the legislature's actions meant that Montana did not have the necessary money to match the $1 million federal appropriation allocated to the state. In 1925, Montana had the dubious honor of being the only state in the union to return Federal Aid money for highways.

The legislature's actions were met with outrage by Good Roads organizations and enthusiasts throughout the state. Meeting at the Placer Hotel in Helena in August 1924, the groups formed the Montana Good Roads Association to develop legislation to establish a state highway fund to match Federal Aid funds. The organization strongly believed that all gasoline tax revenues should go into the fund. The organization was active throughout the state in 1925, raising support for legislation that would enact a three-cent-per-gallon gas tax with the proceeds going directly into the state highway fund. Largely through the association's efforts, the legislature placed Initiative No. 31, the Good Roads Bill, before Montana voters in November 1926.

But until the election occurred, the highway department had to cut back its operations. In April 1925, the department's activities were "confined primarily to the completion of those projects now under contract and to the maintenance of completed projects." The highway commission laid off twenty employees but promised that their jobs would be open to them should conditions improve. Governor Erickson appointed three new commissioners in 1925 and handed over to them the deteriorating situation in the highway department. The new commissioners demanded "rigid economy" by the remaining employees, threatening termination for any "extravagant expenditures," including unnecessary telegrams and telephone calls. The commission closed down the highway department's shop in Helena and sold off its remaining war surplus equipment.[28]

Instead of planning a construction season in 1925, the highway commission concentrated its efforts on the maintenance of highways that had already been improved. Maintenance of the state's primary highway system concerned the BPR and the highway commission. The 1921 Federal Aid Road Act addressed the issue of maintaining the Federal Aid Highway system, which it saw as the protection of its investment in the national transportation system. The federal government, though, was not willing to pay for that maintenance. Thus, when the highway commission created a Maintenance Division in 1921, all of its activities were paid for by non-federal dollars. Between 1921 and 1927, the counties reimbursed the state for maintenance done by the highway department on Federal Aid roads. The counties had difficulties raising money to pay the costs, which rose each year because the amount of constructed mileage increased. In 1929, Mary Bean, wife of Carbon County farmer Jordan Bean, wrote a scathing letter to Governor Erickson about the Montana Highway Commission and the condition of the state's roads:

> *I have just returned from a trip thru Idaho, Utah, Colorado, and Wyoming, and after our experience with the State Highway Commission of Montana—we paid particular attention to Roads. If we hadn't our attention would have been called to them by other tourists—when they noticed our license plates they would laugh and say, "Deliver me from Montana Roads and high gas," and the worst of it we had no "Come back."*[29]

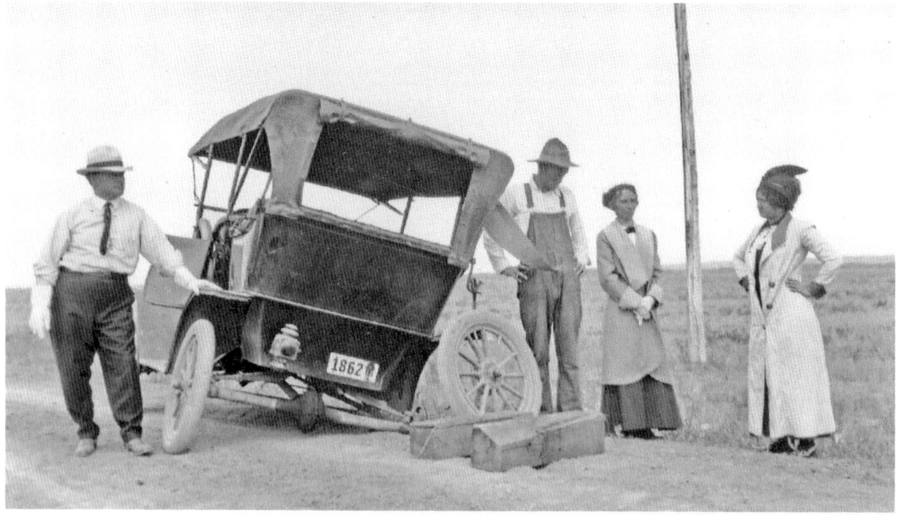

Despite improved roads, flat tires and unhappy passengers were not uncommon on Montana's highways. *MHS Photograph Archives, Helena, PAc 76-26.665.*

Obviously, the existing maintenance system was not fulfilling the goals of the BPR and the highway commission. A portion of the revenue derived from a gasoline tax would go to maintenance of the system.

In 1925, Governor Erickson appointed three new highway commissioners. The democratic-controlled state legislature then reorganized the commission. The three-man panel still awarded contracts and determined policy, but all its members were compensated per diem rather than by salary. The commission also turned over most of its day-to-day responsibilities for the department to the chief highway engineer. Notably, Governor Dixon appointed *Great Falls Tribune* publisher Oliver S. Warden to the commission. A vocal advocate of good roads and well aware of the problems the highway commission faced, he worked hard to change the way the commission and department did business. Largely because of his efforts (which he often paid out of his own pocket), his fellow commissioners appointed him chairman in 1927, a post he would hold until his resignation from the commission in 1935.

The Good Roads Law

In November 1926, Montanans voted on Initiative No. 31, the Good Roads Bill. Aggressively promoted by state good roads advocates, such as the Montana Good Roads Association and the Montana Automobile Association, the state's citizens passed it by a margin of 72,531 votes, demonstrating the importance of good roads to the Treasure State's residents. The initiative provided for a three-cent-per-gallon tax on gasoline, with the revenue going directly into the state highway fund. The highway commission used the revenue to match federal funds for road construction, thereby relieving the counties of that responsibility. The tax placed the burden of road construction on the users of the highways rather than on property owners as it had previously. The increasing number of automobiles on the state's roads meant a great amount of taxes collected through license fees and the gasoline tax, which would ensure a strong financial base for the highway department's construction programs.

The passage of the initiative had a profound impact on the state highway commission: it not only gave the commission and the highway department operating funds but also gave them complete control of road building on the Federal Aid system. As a result of the new law, the commission allocated funds derived from the gasoline tax to each of the twelve construction

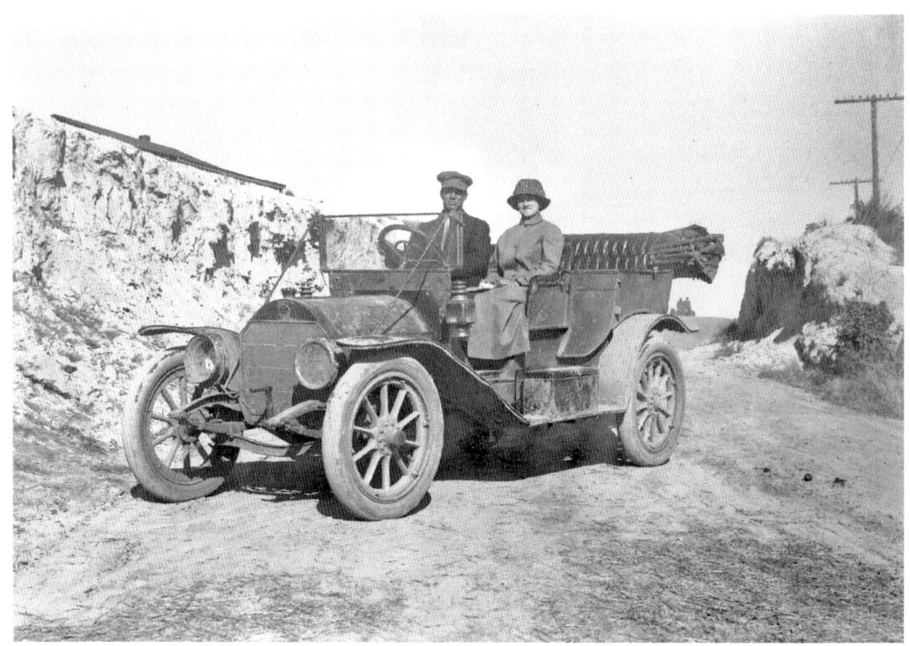

A weekend outing near Polson. *MHS Photograph Archives, Helena, 957-696.*

districts based on the amount of mileage of Federal Aid highway still to be built. The initiative specified that no more than 8 percent of the total cost per project could be spent on administration and engineering. Chief highway engineer Howard Holmes projected that with the gas tax revenue, the department could complete Montana's Federal Aid system by 1937.

Revenue from the gas tax became available to the highway commission in March 1927. Between that date and July 1928, the commission awarded contracts for 121 projects, for a total of just over $2 million. It constructed and improved 544 miles of highway on the Federal Aid system and funded 218 miles of highways on federally owned land. Projects included a new bridge over the Flathead River at Polson, the Scenic Bridge near Alberton and road construction projects in all fifty-six counties. Most of the work was conducted in northeastern and western Montana and consisted of road improvement and surfacing projects. The 1927–28 construction program was the largest in the department's history up to that time.

There was considerable discussion among the highway commissioners, the chief highway engineer and local residents about whether U.S. Highway 93 in Ravalli County should be on the east side of the Bitterroot River or on the west side of the river. In August 1928, the commission, with the BPR's concurrence, routed Highway 93 onto the abandoned Northern Pacific Railway grade on the west side of the river. Other discussions involved channeling U.S. Highway 87 either through or around Fort Benton; creating a connector road from Rockvale in Carbon County south to Red Lodge; establishing the route of the Vigilante Trail (now Montana 287) in Madison County; and switching U.S. Highway 89 between Emigrant and Gardiner from the west side of the Yellowstone River to the east side of the river. The Montana Highway Department implemented a program to build more direct road alignments between the state's communities.

But not all of the department's plans were without controversy. In February 1927, the highway department's engineers and the highway commissioners discussed a plan proposed by Lewis & Clark Caverns owner Dan Morrison and the Morrison Cave Development Association to reroute U.S. Highway 10 through the Jefferson Canyon, thereby abandoning the old Yellowstone Trail route between Cardwell through Jefferson Island to Willow Creek. The new route, while more expensive, would provide a more direct connection between Whitehall and Three Forks and would take advantage of the proximity to the caverns by affording better access for tourists to the natural wonder. The department's administrators, however, believed that the old route was the better one, even though it was three miles longer. The engineers claimed that by keeping U.S. 10 where it was, it could avoid right of way conflicts with the Northern Pacific Railway in the canyon, thereby making the road cheaper to construct.

The commission's decision was not a popular one with area businessmen and tourism promoters. By the first week of June 1927, the commissioners had received the first of many telegrams protesting their decision. The Missoula Chamber of Commerce and the Joint County Affairs Committee of Gallatin County protested the existing routing of the highway and "urged further consideration of the route thru [*sic*] the Jefferson Canyon." The commissioners also heard objections to their decision by delegations from Whitehall, Bozeman and Three Forks over the next months. The delegates all argued that the proposed route through the canyon would "make Lewis & Clark Cavern directly accessible to persons traveling on the main highway and will be more free of snow in the winter time." Ike Pace of the Morrison Cave Development Association wrote to Governor Erickson in 1927 that unless the highway commission chose the Jefferson Canyon route, it would regret the "bitter animosities that are going to be aroused if [it] ignores these popular demands." The highway commissioners stuck to their original argument but hiked the price of the new road up to $150,000 for construction. They did, however, promise to survey both routes before they made a final decision.[30]

While delegations from Gallatin and Jefferson Counties supported the canyon route, the Madison County commissioners backed the existing route via Antelope Creek because it crossed their county. The most vocal supporter of the Jefferson Canyon proposal first appeared at the highway commission's December 1927 meeting. Shadan "Dan" LaHood owned a small wayside store on the original U.S. 10 route but had recently purchased property at the head of the canyon and planned to build a hotel and restaurant there once the new road was constructed. The debate over the routes persisted through 1927 and into 1928, with the highway commissioners receiving petitions and hearing delegations from those in favor of the Jefferson Canyon route and those opposed to it. After the commission's talks with the Northern Pacific Railway proved successful in regards to easements, it opted to construct the road through the canyon in February 1928 and awarded contracts to the Leo T. Lawler Company of Butte for the construction of the road in June of that year.

In addition to Jefferson Canyon, the Good Roads Law enabled the highway commission to tackle many projects it had long planned but did not have funds to accomplish. In 1929, the commissioners awarded the project to the Missouri Valley Bridge and Iron Company to construct the long-awaited bridge across the Missouri River near Wolf Point. Described by the *Wolf Point Herald* as a "dream come true" when dedicated in July 1930, the bridge, which has the longest truss span in the state at four hundred feet, still stands alongside Montana Highway 13. Other projects included a road through Bad Rock Canyon on U.S. 2 east of Columbia Falls and a study for a new bridge across the Missouri River north of Lewistown. For four years between 1927 and 1931, the highway commission, BPR and National Park Service hashed out the details for a bridge across the Yellowstone River at Gardiner, the north entrance to Yellowstone National Park.

The 1920s were also an active time for the Park Service in regards to road projects. Unlike the highway commission, which strove to build roads to facilitate private and commercial traffic, the Park Service intended its roads to encourage visitation to the national parks. In 1925, the Park Service began construction on one of the most spectacular roads in the world: the Going to the Sun Highway in Glacier National Park. Conceived by government engineers in 1917 as a way to increase visitation to the park, it wasn't until 1924 that surveyors mapped out a route for the highway across the 6,646-foot Logan Pass. Supplies, including explosives, had to be carried to the

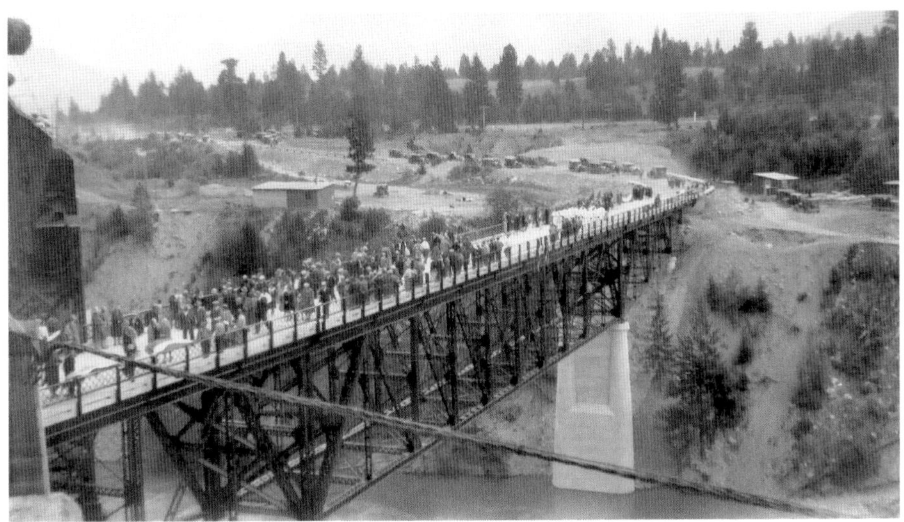

The dedication of the Scenic Bridge spanning the Clark Fork River east of Superior in May 1928. *MDT.*

construction site on the backs of mules and horses until a road wide enough to accommodate heavy equipment had been carved, scaled and blasted up the mountains to the pass. Although motorists could reach the summit of Logan Pass by 1929, it wasn't until July 1933 that the $2.5 million road was opened between West Glacier and St. Mary. One of the engineering wonders of the world, the Going to the Sun Highway continues to thrill park visitors with its scenic vistas and stomach-turning, cliff-hugging switchbacks.

The Montana Highway Department's ambitious road program began in 1922 and continued throughout the decade. The BPR implemented a phased construction program in 1922 that it utilized into the 1930s. It directed that the state highway departments first grade and drain roads on the Federal Aid system. At a later date, they could put a better surface on the roads after the sub-grade had a chance to stabilize. The BPR concentrated the program in the West, the prairie states and the South, where traffic was light "and roadbuilding [*sic*] least advanced." The system allowed for the rapid improvement of the Federal Aid system.[31]

Gravel was the standard road surfacing material in Montana. In eastern Montana, contractors placed scoria on roads as it provided a better covering to the native soil, which turned to gumbo when wet. The highway department paved with concrete only 40 miles of highway during the 1920s. Early in 1920, chief highway engineer John Edy agreed, on behalf of the highway commission, to purchase Portland cement from the Three Forks Portland Cement Company. In March 1920, the commission awarded a contract to M.J. Medin & Company to pave 9.11 miles of the highway between Butte and Anaconda, the first concrete-paved highway in Montana.

In January 1929, the highway commissioners discovered that they had a big problem to contend with. The gasoline tax applied to all fuel sales and did not make the distinction between on-road use and use by farmers and ranchers for their off-road operations. The Montana State Board of Equalization annually issued refunds to farmers and ranchers to compensate them for their payment of the tax for non-highway use. The board notified the highway commissioners that the refunds for 1927 and 1928 far surpassed the anticipated revenue they would derive from the gas tax. The commissioners calculated an $188,000 shortfall. Without the gas tax funds, the state could not provide matching funds for federal aid highway dollars and was unable to plan for a construction program in 1929. The commission asked Governor Erickson to find money so it could make payroll in January and February. A feeling of panic permeated the highway commission meetings during the first three months of 1929 as it attempted to find solutions to the problem. The commissioners asked the attorney general if it would be possible to obtain warrants on future gas tax revenue. Lewis Ketter of the attorney general's office determined that it could not legally do so. The highway department responded to the funding crunch by terminating most of its employees and reducing its operations to the "greatest possible extent."[32]

Eventually, the legislature came to the rescue. Through House Bill No. 94, it enacted a two-cent surcharge on the gas tax to provide the necessary money. The legislature also passed a bill in March 1929 that allowed the commission to transfer money from the highway fund to a highway trust fund in case of future emergencies. Unfortunately, the 1929 problem proved to be a foreshadowing of what was to come when the Great Depression hit Montana in full force the following year.

Federal and state legislation in the 1920s established the basis for highways and bridge construction in Montana for the next three decades. For much of the '20s, the Bureau of Public Roads strove to take financial responsibility away from the counties and give it to the Montana State Highway Commission and the Montana Highway Department. This financing, standardization and professionalization of the highway construction industry in Montana caused a profound change in how roads and bridges were built in the state. The new system established the state/federal matching formula and the 7 Percent System in 1921. The state enacted a gasoline tax to match federal

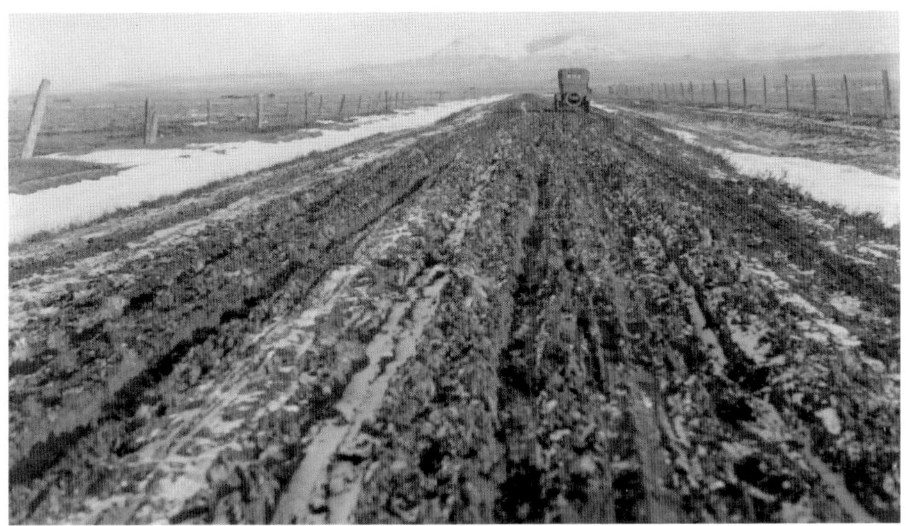

Treacherous roads during the winter made travel between Big Sandy and Havre difficult in December 1922. *MDT.*

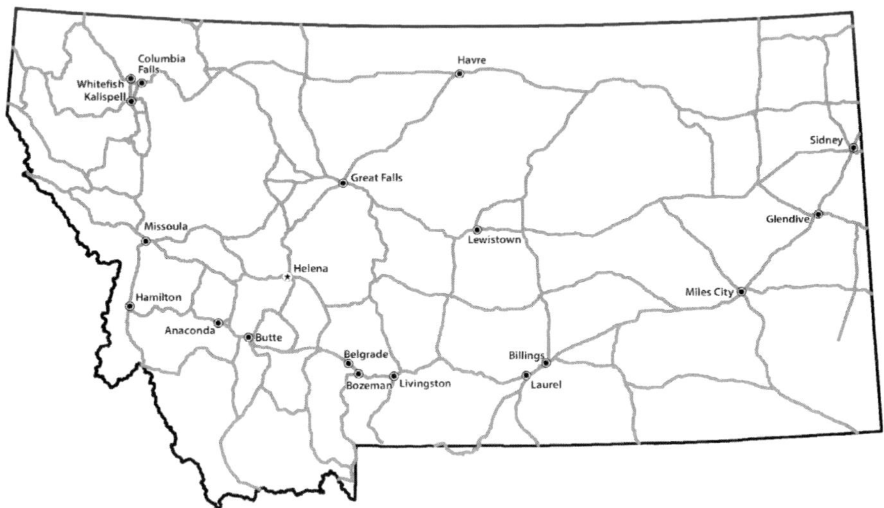

Montana's Federal Aid Highway System, 1941.

money and systematized the road-building program in 1927. The result was the most prolific period in Montana road building since the 1860s. The highway commission oversaw the improvement of 2,448 miles of Federal Aid highways and spent nearly $28 million doing it. It was the beginning of the modernization of Montana's highway system.

Part IV

TRANSFORMING MONTANA'S HIGHWAYS DURING THE DIRTY THIRTIES

Years ago wild game and Indians made the trails in Montana. Now high-powered engineers and contractors are responsible. These scientific gents have built about 5,000 miles of oiled arterial system and draped it over the landscape where it will do the most good. At that they are using the mountain passes the old timers located. These new highways are safe, direct, and dustless. You will enjoy driving them.
—Montana: High, Wide and Handsome, *1940*

The flush times for road building in Montana during the late '20s came to a grinding halt at the beginning of the 1930s. The Great Depression devastated Montana. But out of that national calamity arose a modernized highway system transformed by increased federal involvement in Montana road building. Montana had among the worst highways in the nation in the 1920s, but within a few years, the state boasted one of the best highway systems in the United States.

While the Great Depression struck Montana in full force in 1931, the state had begun to feel the effects of it long before then. Drought struck eastern Montana in 1929 and grew in intensity over the following years. Although there had been a period of brief prosperity in the late 1920s, farmers and ranchers, still recovering from the post–World War I drought and depression, proved unable to cope with a deeper economic decline that began in 1930. Drought and grasshoppers destroyed crops for which market prices were already low. Meat and wool prices collapsed, ravaging Montana's ranchers.

The debt-ridden Anaconda Copper Mining Company, overextended from overseas investments, could not compete with cheaper copper flooding the market from South America and South Africa. Copper prices fell, and mining and smelting declined, causing layoffs in the state's industrial belt. The ripple effect was felt in the timber industry as well. Montana's unemployment rate had swelled to 22 percent by 1933. Destitute families relied on county governments and private relief groups for assistance, all of which were hopelessly strained and unable to provide much aid.

The economic disaster hit the Montana Highway Department hard. It depended on license and gasoline tax revenues to provide matching funds for federal allocations for highway construction. Those revenues steadily declined, making it difficult for the highway commissioners to fund new construction projects. With no matching funds, there would be no federal money to build roads and keep contractors employed. Increasingly, county and city delegations appeared before the highway commission desperate for highway work in their districts. Highway projects meant better roads and new bridges but also, most importantly, jobs for the unemployed. While the highway commission sympathized with the delegations, there was little it could do to help them.

In 1930, the highway commission awarded just $2.7 million in contracts, a more than 30 percent drop from the money spent the previous year. Contracts consisted primarily of grading and road surfacing projects and the construction of bridges, including structures over the Yellowstone River near Reed Point and the Missouri River south of Cascade. Most of the projects concentrated on U.S. Highways 10 and 91, with others on U.S. 2 and what is now Montana Highway 200. The highway department established a uniform width of eighteen feet for roads and adopted standard route markers to designate them. In addition, the highway commissioners implemented a wage scale—its first—for projects only in the Butte area: $5.00 per day for common laborers and up to $5.50 for grader operators. The state legislature adopted a similar wage scale in 1931 for all highway projects no matter where they were located. The seemingly busy construction program in 1930, however, could not prevent continued departmental layoffs as revenues continued to dwindle.

Nationally, the slowdown in highway construction and unemployment were major concerns of President Herbert Hoover and Congress. At the request of the president, Congress passed the Emergency Construction Act of December 20, 1930. The legislation allocated $80 million for Federal Aid highway projects, along with $3 million for projects on federally owned and

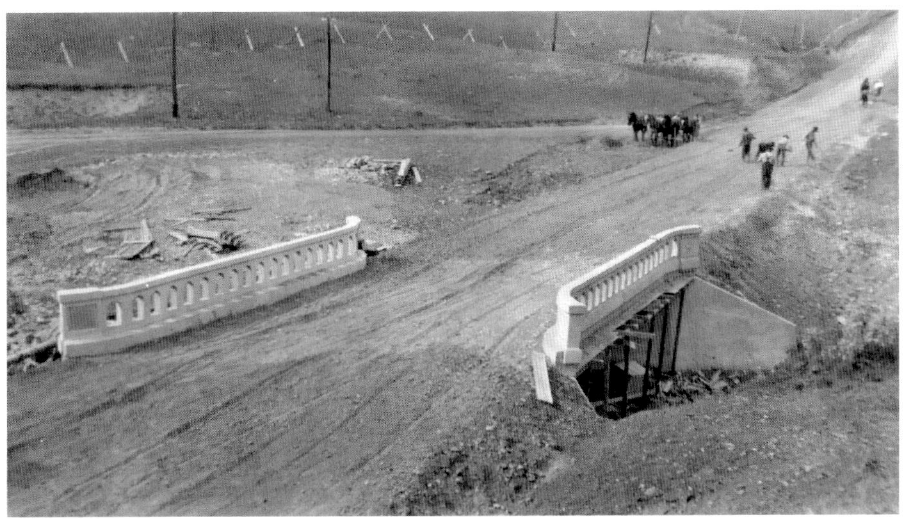

The Dearborn River Bridge nears completion on U.S. 287, June 1929. *MDT.*

administered lands. The Oddie-Colton Act of 1930 provided for 100 percent federal funding on roads built across Indian reservations and public domain in order to speed up construction. Prior to 1930, the states' participation in public lands projects was minimal, but now the states would be responsible for the design and construction of roads across those lands. The December 1930 appropriation was not a federal grant to the states but a loan to be repaid out of future federal aid apportionments. The states used the money to match funds provided by the Federal Aid Road Act of 1930. In addition, the money had to be obligated for work completed by September 1, 1931.

The Montana State Highway Commission and Montana Highway Department received word of the impending legislation in November and was well prepared when the state received its $1.6 million share of the emergency funds. The highway department's engineers had already completed the necessary surveys and construction plans for fifteen road and bridge projects. In January 1931, the commissioners awarded the first fifteen Emergency Federal Aid Project contracts, following it up in February and March, letting thirty-six more projects. In all, the highway department was able to improve over one thousand miles of Montana's primary highway system under the legislation.

While most of the Emergency Federal Aid Projects consisted of simple grading and surfacing projects, there were a few major construction projects as well. The most visible of these was the construction of a direct highway route between Helena and Great Falls through the rugged Prickly Pear and Missouri River Canyons. In the late 1920s, there was no road between Wolf Creek and Cascade. For motorists, the old Benton Road was still the only way of reaching Great Falls from Helena. Beginning in 1928, however, delegations from Helena, Great Falls and Cascade petitioned the highway commissioners to construct a more direct route between the cities. Their lobbying proved successful, and the commissioners designated a new route between Helena and Great Falls as a component of U.S. Highway 91. Over the course of the next three years, the highway commissioners awarded seven contracts for the construction of the road—all funded by emergency federal and state relief money.

The narrow Prickly Pear Canyon had limited the options for road alignments since the 1860s. The Great Northern Railway's Montana Central Railroad, built in 1887, occupied the best route through the canyon with a county road closely paralleling it. In order to build the road to the standards of the day, the highway commission coordinated its efforts with the railroad. The railroad's engineer did not get along with the contractor, Sam Orino, and the man was, on one occasion, caught red-handed stealing fill material from the contractor. The location of the community of Wolf Creek at the mouth of the canyon also posed a significant problem. The engineers bypassed the town, which caused many of its businesses to relocate to the new highway. The new road also required considerable blasting through the canyon's walls. Despite the difficulties, Orino managed to carve a modern scenic highway through the canyon, but not without complications between him and the highway commissioners. During the course of the project, Orino managed to violate most of the provisions of the federal relief acts in the hiring and paying of laborers.

The route north of Wolf Creek through the Missouri River Canyon proved much easier for the Morrison-Knudsen Company to build. Constructed between 1930 and 1933, the area was sparsely settled. Little blasting was necessary through the river canyon, and the road, perhaps more than other routes in the state, appeared to be "draped" over the landscape. The project included two major steel bridges over the Missouri River northeast of Wolf Creek and at the community of Hardy. Because of the scenic nature of the route, moreover, the bridge engineers built reinforced concrete bridges rather than the standard timber bridges to span the numerous creeks between Wolf Creek and Cascade. When completed in 1933, the forty-mile Sieben-Cascade section of U.S. Highway 91 was one of the most picturesque in the state, with plenty to offer in the way of recreational opportunities for motorists.

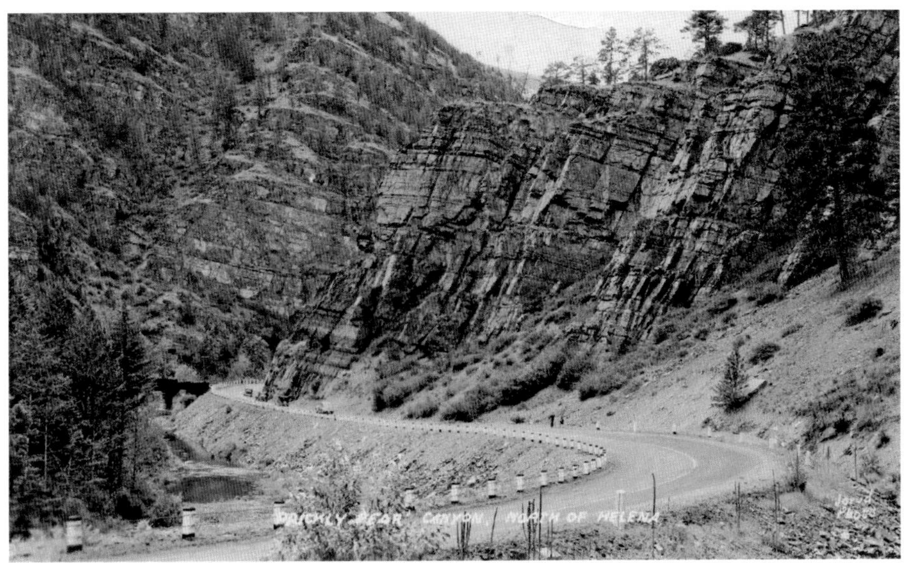

U.S. Highway 91 in Prickly Pear Canyon, north of Helena, in the early 1940s. *MDT.*

Shortly after the legislature convened in January 1931, Governor Erickson, Attorney General LeRoy Foot, the state Board of Examiners, the highway commissioners and several Good Roads organizations successfully petitioned the lawmakers to approve an unsecured bond called a debenture to provide matching money for federal funds. Erickson signed House Bill No. 1, the Gasoline Tax Debenture Bill, into law on January 26. Rechristened the Highway Treasury Anticipation Debenture Act by the highway commission, it raised $1.5 million each year for highway construction from 1931 to 1934. Within weeks of its passage, however, the Montana Supreme Court nullified the debenture on the grounds that it obligated the state for more than the $100,000 liability specified in the state's constitution. The $100,000 could be exceeded only if authorized at a general election. The referendum election, held in May 1931, carried forty-eight of the state's fifty-six counties. After surviving a legal challenge brought against the highway commission by two Augusta-area residents, the state sold the debenture's securities to a syndicate of Montana banks. The debenture gave Montana money to match its regular federal apportionment along with additional federal emergency appropriations. Montana now had more money available for the 1931–32 construction seasons than it did in the 1929–30 seasons—despite a significant drop in vehicular and gas tax revenues.

By the summer of 1931, Montana and out-of-state contractors employed three thousand Montanans on projects throughout much of the state. Federal legislation mandated that contractors rely primarily on pick and shovel work with only a minimal amount of heavy machinery. The regulations also obligated contractors to hire local residents and pay those who could also provide horses to pull scrapers and other equipment at a higher rate than common labor. The department was unable to provide much work to drought-ravaged northeastern Montana other than a single grading project on U.S. Highway 2 between Culbertson and the North Dakota line.

Even with the enormous infusion of federal relief funds into Montana in 1930 and 1931, it did not significantly lessen the impacts of the Great Depression. Federal rules limited the money to highway projects, and it was not used for the general relief of thousands of Montanans who were in dire need of help. The pervasive drought in eastern Montana and the nascent copper and timber industries in western Montana, along with declining tax revenues, impeded the state's efforts to get back on its feet. The state relied heavily on federal emergency relief apportionments to keep the road builders active.

Importantly, the Federal Emergency and Construction Act included stipulations regarding minimum wages and the number of hours laborers could work each week. Although the state already had its own wage scale law, the act set the minimum wage for unskilled labor at $0.50 an hour, with skilled

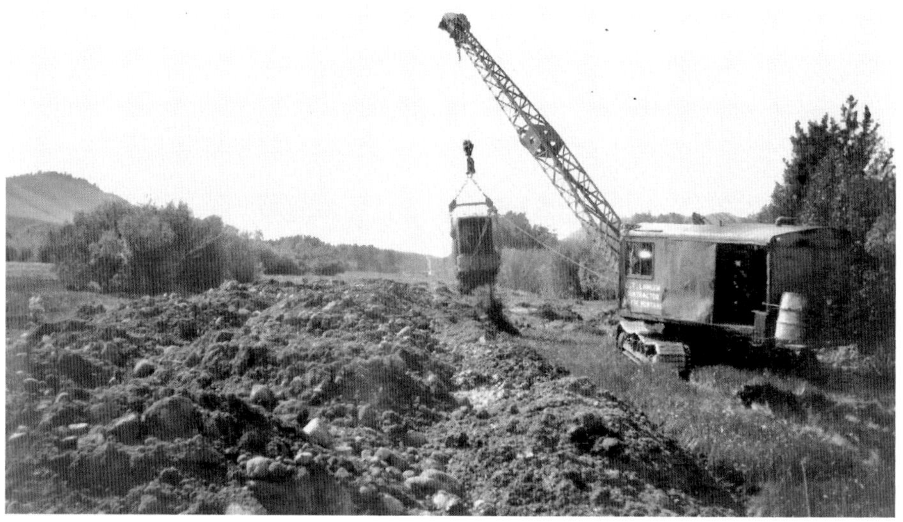

Reconstructing U.S. Highway 287 north of Ennis, 1932. *MDT.*

labor at a maximum of $5.50 per hour. The legislation stipulated thirty-hour workweeks and employment preference for veterans with dependents. The Montana law matched closely the federal law. The state legislature amended the Montana law to include veterans in 1933. Highway projects in remote areas of Montana required that contractors maintain "labor camps" for their workers. Federal law mandated that contractors provide one-dollar-per-day room and board per worker in those areas. On those projects, contractors suspended the thirty-hour workweek and implemented a six-day workweek to speed the completion of projects. The camps consisted of nothing more than collections of tents. Though isolated, people not on the payroll often stopped at the camps for free meals.

Many of the road projects in the early '30s involved placing a rudimentary paved surface, called oiling, on the state's roads. Road oil was, essentially, an asphaltic treatment of the road grade that produced a type of blacktop surfacing. With oil refined by local refineries, contractors used three methods to oil roadway surfaces. The cheaper method involved placing light asphaltic oil on a compacted gravel road surface and allowing it to penetrate the gravel. Contractors then placed heavier asphaltic oil over that and covered it with stone chips or fine gravel. The road mix method of oil surfacing involved mixing, in place, asphaltic oil with crushed rock or gravel by blade graders. The most expensive method of treatment was the plant mix method. It involved mixing asphaltic oil and crushed rock or gravel in central mixing plants. The materials were then trucked to the project site and deposited on the road. Graders shaped the surfacing, which was then compacted by steamrollers. The plant mix method had the advantage of allowing a longer work season and required less maintenance.

The Beartooth Highway

By far the most ambitious highway project in the early '30s was the construction of a road over the Beartooth Plateau between Red Lodge, Montana, and Cody, Wyoming. It was not, however, the first attempt to build a highway from Red Lodge across the mountains to the northeast entrance of Yellowstone National Park. The Black and White Trail was the first highway project let under the auspices of the Montana State Highway Commission in 1919. The brainchild of Red Lodge physician J.C.F. Siegfriedt, local promoters hoped the route would stimulate tourism in the area. Construction crews completed only a portion of the route, including thirteen switchbacks, when the highway commission abandoned the project.

A few years later, a group of Red Lodge businessmen lobbied Congress to construct a new road between Cooke City and Red Lodge, whose coal mines had recently closed. Hoping to revitalize the depressed community through tourism, the businessmen sent *Red Lodge Picket* publisher O.H.P. Shelley to Washington, D.C., to garner support for the project. By the late 1920s, he had gained the backing of Montana's congressional delegation to build the project. The delegation, all good roads advocates, sponsored a bill to create the Park Approach Act, which President Hoover signed into law on January 31, 1931. The act specified the construction of scenic routes to the country's national parks through federally owned land. The Beartooth Highway was the only road constructed under the Park Approach Act.

In June 1931, the federal Bureau of Public Roads (BPR) hired five companies to build the highway. The Morrison-Knudsen Company obtained the contract to build the road on the more demanding Montana side of the border. Construction on the $2.5 million project began in 1932 with the BPR supervising the work. Although an approximate route for the road had been surveyed in 1925, the Beartooth Highway is an excellent example of "seat-of-your-pants" construction, with many of the engineering decisions made in the field. The contractor employed around one hundred workers to blast and carve their way up the side of the eleven-thousand-foot plateau. The workmen gave names to many features of the road that are still used today, including Lunch Meadow, Mae West Curve and High Lonesome Ridge. Two men died as a result of accidents during construction. The workers lived in tent camps at the base of the plateau on both the Montana and Wyoming sides of the project. By the summer of 1935, sufficient work had been completed on the four switchbacks, allowing travelers to use the road to reach Yellowstone Park and Red Lodge for the first time. The road officially opened on June 14, 1936. Although declared impossible by many engineers of the day, the Beartooth Highway remains a singularly spectacular road that has no equal in the lower forty-eight states.

By the end of the Hoover administration in 1932, Montana's Federal Aid highway system consisted of 5,012 miles. Of that mileage, the highway department had improved 3,985 miles of road, including permanent alignments and grading. Most of that mileage had been surfaced with gravel, rock or scoria; some had been oiled; and only 51 miles were concrete paved. The state highway commission had spent a little over $16 million on Montana's highways since 1930, yet tax revenues continued

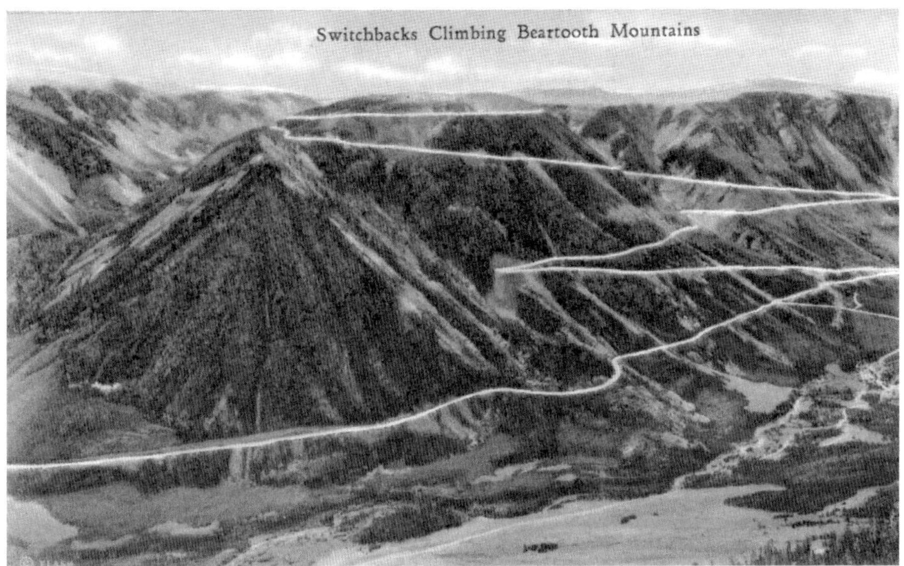

The Beartooth Highway provided access to Yellowstone National Park and was a draw for tourists after its completion in 1936. *Author's collection.*

to dwindle and unemployment worsen. The federal government seemed increasingly unwilling to deal with the crisis. By the time of the November 1932 presidential election, the highway commission was unsure if it could maintain any kind of a highway construction program.

The New Deal

With his reputation shattered by three years of depression and his apparent callousness toward those most impacted by it, Herbert Hoover had little chance of reelection in 1932. Franklin Delano Roosevelt ran an optimistic campaign promising a New Deal for the American people. That confidence proved infectious for Americans, and FDR won the election by a landslide in November 1932. Montana voters came out overwhelmingly for Roosevelt. His election inaugurated the greatest reform movement in U.S. history and permanently changed the relationship between federal and state governments and, significantly, between the federal government and the people. Roosevelt's New Deal came to Montana in the nick of time.

The eight-year period from March 1933 to December 1941 marked a profound change in how the Montana Highway Department financed, built and maintained roads and bridges. During that time, the department reconstructed or upgraded 6,563 miles of Federal Aid highway, including 1,539 miles of bituminous surfacing and 2,996 miles of oil-surfaced highway. New Deal programs included the establishment of a feeder or secondary road program to improve approximately 723 miles of farm-to-market roads that were not on the Federal Aid system. The improvement of Montana's highways was a lifesaver for thousands of the state's unemployed as federal government relief programs put them to work on highway construction projects. As a result, Montana's highways transformed from among the nation's worst highway systems into one of the nation's best within a just a few years. A tourist from New York wrote to Governor Frank Cooney in November 1935 that his "impressions of the State highway systems were varied, but in reference to Montana that none of the States, regardless of size, population, or resources, gave me quite the satisfaction that your state—Montana—gave me."[33]

Shortly after his inauguration, FDR pushed through the National Industrial Recovery Act (NIRA), the first sweeping federal legislation to combat the effects of the Great Depression on the nation's unemployed. Title I of the act appropriated $400 million for the construction of public highways. Unlike the usual Federal Aid Road Act, the money constituted a federal grant to the states, which did not need to provide matching funds. Importantly, FDR intended the legislation to provide work for the thousands of unemployed by putting them to work on highway projects. The NIRA set minimum wage scales for unskilled and skilled labor, gave hiring preferences for World War I veterans and local labor, stipulated a thirty-hour workweek and focused on pick and shovel work rather than the use of machinery. The act required that 25 percent of each state's apportionment be spent on farm-to-market roads, 25 percent on municipal roads and 50 percent on Federal Aid highways (forest highways included). The formula for which National Recovery Highway (NRH) money was allocated to the states was based on the ratio of population to the amount of public land in each state. Because Montana was a "public lands state," it received a higher federal appropriation under NIRA than non-public lands states.

The highway commissioners awarded the first NRH-funded projects in August 1933. At just over $1.6 million, it was the largest single contract letting in the commission's history. The sixteen projects included the grading and surfacing of just over 147 miles of the state's primary highways and

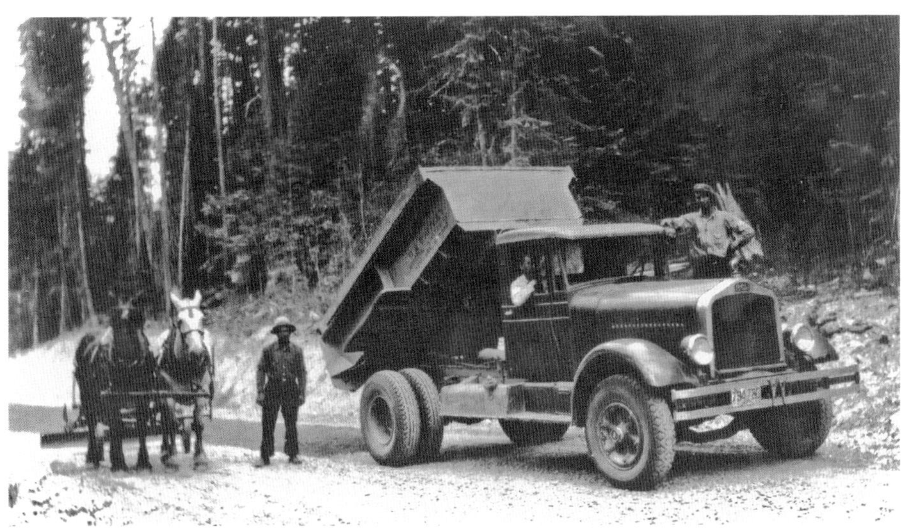

Contractors relied on a mixture of horse-drawn and power equipment on U.S. Highway 10 in western Montana. *MDT.*

the construction of 51 bridges, including the much-anticipated Missouri River Bridge at Culbertson. Over the next twenty-two months, the highway commission awarded 229 contracts for over $12 million. In all, contractors graded and surfaced, either with oil or bituminous pavement, 721 miles of primary highway in the Treasure State. The NRH also funded the construction of 237 bridges and 322 miles of secondary highways. The commission let the first "feeder" road contracts in November 1933 for projects near Stanford, Malta and Harlowton.

For the first three years of the New Deal, the federal government waived the usual Federal Aid Road Act prerequisites for relief legislation that required state matching money. In June 1934, Congress supplemented New Deal funds by implementing the Hayden-Cartwright Act, which provided an additional $3.7 million to Montana for road and bridge construction. In May 1935, the U.S. Supreme Court declared the NIRA unconstitutional because it concentrated too much power in the hands of the executive branch. To replace it, FDR pushed through Congress the Emergency Relief Appropriation Act of 1935. In 1936, Congress reinstituted the regular Federal Aid apportionments, which caused the Montana State Highway Commission considerable problems until the beginning of World War II. Despite a slowly improving economy, the state was still unable to raise the necessary matching funds from gasoline tax revenues. Instead,

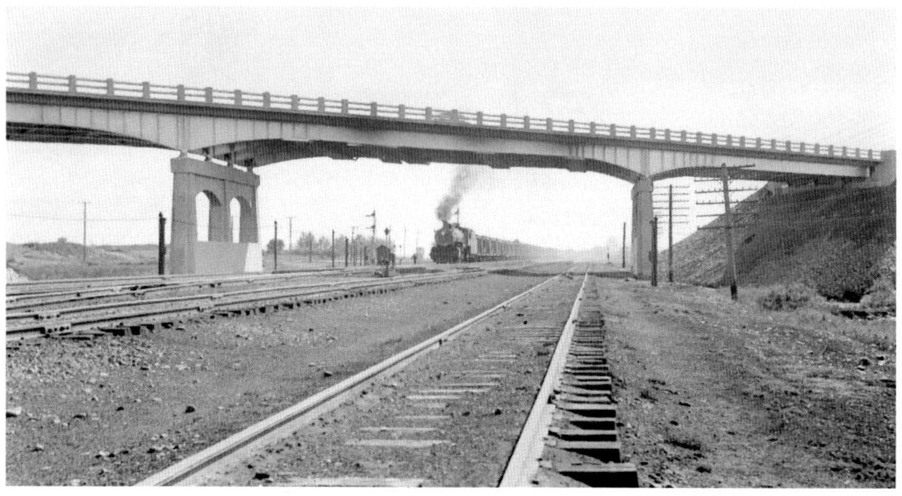

The Mossmain Overpass near Laurel was one of forty railroad-grade separation structures built by the Montana Highway Department during the Great Depression. *MDT.*

the state relied on debentures to provide the necessary matching money for federal apportionments.

The Emergency Relief Appropriation Act continued many of the provisions of the NIRA and established the Works Progress Administration (WPA), one of the most successful federal programs in U.S. history. Through the WPA, Montana continued the highway-building boom it had enjoyed since 1933. Importantly, the act appropriated funds to correct or eliminate dangerous at-grade railroad crossings by building grade separation structures. The Works Progress Grade Crossing Highway Program proved to be one of the most successful of the New Deal. Between 1936 and 1941, the highway department designed and constructed forty railroad overpasses that eliminated dangerous at-grade railroad crossings on heavily traveled routes. Like other federal legislation during the boom years of the Great Depression, the Appropriation Act constituted a grant to the states and did not require matching money for the federal apportionments. The grant money went directly to federal relief and "make work" agencies.

At the time of the New Deal legislation, Montana's Federal Aid system consisted of 5,497 miles, of which 1,080 miles were still in need of improvement. By November 1934, the highway commission had placed under contract 60 percent of the National Recovery funds at its disposal, with 80 percent committed by January 1935. The highway commission awarded contracts for the remaining 20 percent in March 1935. The Agricultural

Howdy Everyone! Glad to See You!

Highways and tourism became inseparable in the automobile age. The trail associations promoted automobile tourism but concentrated on getting motorists to the national parks. The associations did little to promote local attractions, instead publishing advertisements for hotels, restaurants, auto camps and garages along the routes. There were few encouragements for tourists to stop and enjoy all that Montana had to offer. Initially, the Montana State Highway Commission did little to support tourism. In 1920, commission chairman Frank Conley categorically stated that the primary function of the state's highway department was for commerce, not tourism. It wasn't until 1926 that the highway commissioners recognized tourism as an important function of Montana's highways, but they did little to encourage it. For the most part during the 1920s, Montana's highways were poor and sometimes barely passable. Only the most courageous automobilists attempted long trips to traverse them, leading author Hoffman Birney to write:

> *The roads of Montana are, I believe, the poorest of any state in the Union. Even the glorious scenery of the Rockies can't entirely make up for ruts, chug-holes, mud and detours—to say nothing of broken springs or stone-bruised tires. I turned off the highway some twenty miles from West Yellowstone, heading northward across Reynolds Pass toward Ennis, Montana, and Virginia City, my goal. The road was atrocious, the scenery superb. Great pine-covered mountains—the Rockies themselves—reared their magnificent heights on either hand.*[34]

Automobile tourism in Montana was not for the faint of heart before the Great Depression.

The golden age of automobile tourism in the state coincided with the nation's greatest economic calamity: the Great Depression. The transformation of the state's highways to a modern system during the '30s made tourism the third most important industry in Montana behind agriculture and mining. Good, paved roads enticed tourists from all over the United States and Canada to come to the Treasure

State, and the Montana Highway Department quickly took advantage of that fact.

In 1934, the highway department published its first user-friendly highway map. Unlike previous maps produced by the department, this one was not intended as a planning tool. Instead, it showed the major routes—paved and unpaved—clearly delineated on a colorful map that could be easily folded to fit in an automobile's glove compartment. The map proved to be popular with tourists and Montana residents. The highway department's graphic artist, Irvin "Shorty" Shope, illuminated the map covers with colorful illustrations advertising Montana's scenic beauty and connection to the Old West.

The map was the idea of the highway department's plans engineer and unofficial tourism director Robert H. Fletcher. Over the next seven years, Fletcher and his small staff formulated Montana's first state-sponsored tourism program. The highway commissioners realized the importance of tourism to the continuing highway construction program: tourists paid a substantial part of the gasoline tax. Consequently, they supported most of the ideas that Fletcher and his staff devised during the decade. The state's tourism program relied on both national publicity and word-of-mouth reports made by visitors to the state. Most of Fletcher's efforts concentrated on keeping visitors in Montana as long as possible by directing them to things to do and see.

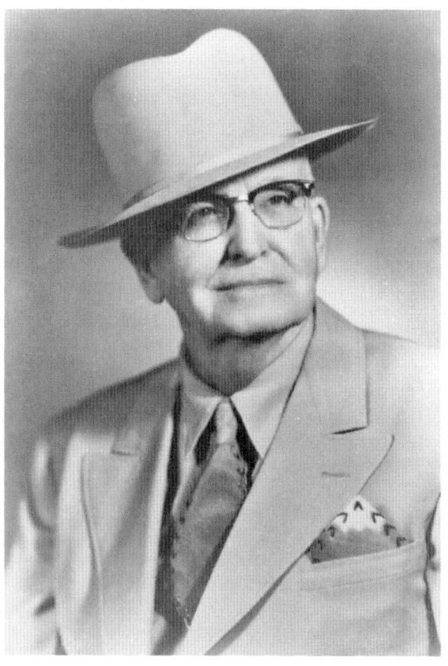

Robert H. Fletcher devised many of the highway department's programs that branded Montana as a tourist destination during the 1930s. *MHS Photograph Archives, Helena, 942-134.*

In May 1935, the highway commissioners provided $5,000 to Fletcher to develop and install thirty roadside historical markers. A natural-born storyteller,

Fletcher wrote the marker texts, and highway department sign shop foreman Ace Kindrick hand lettered the text onto five- by eight-foot plywood sheets. Fletcher intended the signs to reflect Montana's Old West heritage. They told stories about Montana's colorful history in a manner that was "pepped up a little for people who are human and don't take themselves and life too seriously." By 1941, the department had installed one hundred roadside historical markers.[35]

The department followed the success of the highway marker program in 1936 by opening nine port-of-entry stations at strategic locations around the state. The stations were for many tourists the first point of contact with Montana. The small stations looked like log cabins, keeping with the overall rustic theme of Montana's tourism programs. Station attendants, all uniformed college-age boys, provided maps, brochures and information to tourists, who were encouraged to stop at the stations when they entered the state. For a time in the late 1930s, station attendants distributed postcards to each out-of-state automobile asking that visitors draw in the routes they had taken across Montana and drop them in the mail along with comments about their stay in the state.

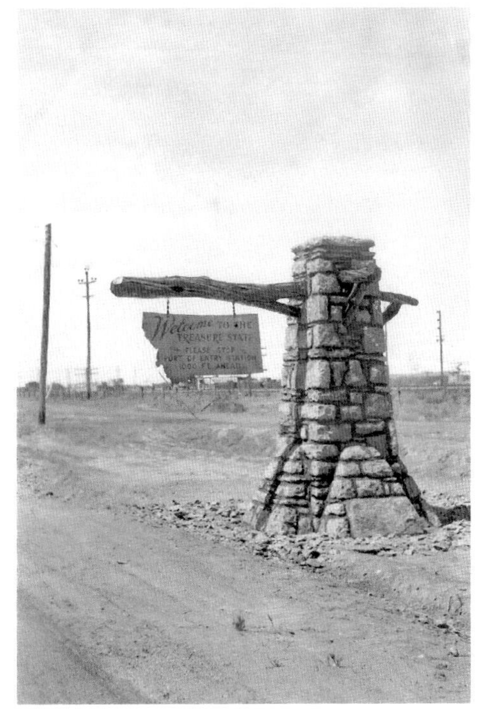

Stone monuments like this one greeted motorists when arriving in Montana after 1935. *MDT.*

Other programs developed by Fletcher included the establishment of roadside parks and picnic areas, tourist information shacks, rest areas, promotional brochures and roadside museums to display artifacts and dioramas describing Montana's history. The first museum, located on Main Street in Laurel, exhibited archaeological artifacts excavated from nearby Pictograph Cave. In 1938, the highway commission purchased Pictograph Cave and made it a tourist attraction. A popular speaker at civic clubs and conventions, Fletcher touted the importance of tourism to Montana's economy. Fletcher resigned his position at the highway department in 1941 to become the public relations director for Montana Power Company.

Fletcher's departure from the department and the nation's entry into World War II temporarily halted the state's tourism program. The highway commission reinitiated it after the war, expanding many of the programs Fletcher developed in the 1930s, including the roadside historical markers, the ports of entry, annual road maps and promotional booklets. The Montana Highway Department was responsible for the state's tourism programs until the 1960s.

Appropriation Act of 1936 reinstituted the regular federal aid appropriation system. Unfortunately, as would be the problem for the rest of the decade, the commission did not have enough money to match the expected Federal Aid appropriation.

Countdown to War

Although World War II began in Europe in September 1939, it was nearly a year before official discussions of it occurred in highway commission meetings. In June 1940, Congress passed the Federal Aid Highway Act of 1940. Similar in content to the 1938 legislation, it included a 25 percent cut in federal appropriations to the states. The states received $160 million for road construction, of which Montana's share was a little over $2.6 million. Importantly, the commissioners also received word that American Association of State Highway Officials (AASHO) president Henry F. Cabell offered FDR the "full services and facilities of the membership of [AASHO] for any services that they can render in any capacity in any undertaking in the present emergency." Three months later, in September, the federal government, through the Public Roads Administration (PRA), asked the states to cooperate in bringing Federal Aid highways up to military roads standards.[36]

That request precipitated heated discussion between the PRA and the highway commission about the establishment of a national Strategic System of Military Defense Highways. The PRA had already prioritized the primary highways it wanted on the system in Montana before it sought the comments of the state highway commission. The PRA categorized highways into three priorities, with Priority One highways being the most important in the event that the nation went to war. In Montana, these included U.S. Highways 10 and 91 and U.S. 87 from the Wyoming border to Billings. Priority Two roads were U.S. Highway 2, Montana Highways 5 and 13 and Secondary 511 in northeastern Montana; U.S. 87 from Billings via Roundup and Lewistown to Great Falls; and U.S. 12 from Missoula to Lolo Pass. The Priority Three highways were U.S. Highway 93 from Missoula to Kalispell and Montana Highways 13 and 200 from Wolf Point via Circle to Glendive. Under the proposed federal legislation, projects scheduled for Strategic Highways received priority in funding and materials.

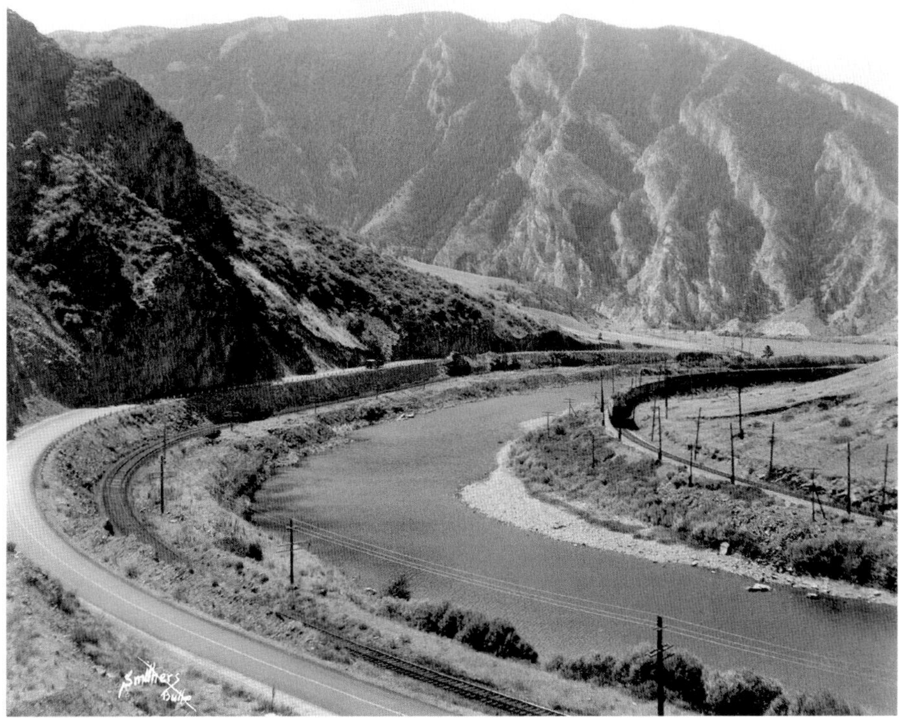

U.S. Highway 10 in Jefferson County represented the transformation of Montana's highways in the 1930s. *MDT.*

Despite the fact that the highway commission awarded over $6 million in contracts in 1941, most of the discussion in its meetings involved strategic highways and the Defense Highway Act of 1941. The act reflected the federal government's growing emphasis on national defense and the prioritization of money spent on strategic highways. The discussions came to a head at the December 8, 1941 meeting, the day following the Japanese attack on Pearl Harbor and the day President Roosevelt asked for a declaration of war against Imperial Japan. The month previous, the highway commission had awarded nearly $1.2 million in contracts to road contractors. The commissioners announced that the contract awards would stand but advised the contractors that it could not guarantee priorities on non-strategic highways, and they should proceed at their own risk. The warning proved prophetic as the federal government cancelled all road contracts, suddenly ending what had been an eleven-year period of the steady and spectacular improvement of Montana's highway system.

Part V
A Time of Unprecedented Construction

The Postwar Years

There has never been a time in the history of the department when so much work has been accomplished in the design and preconstruction departments with so few men.
—A.F. Winkler, chairman of the Montana State Highway Commission, 1945

War and peace had a significant impact on the Montana Highway Department between 1942 and 1956. During the war, many employees left the department to serve in the armed forces or work in the war industries. Because there was no Federal Aid forthcoming to the state for roadwork during the national emergency, the highway system the department worked so hard to build in the 1930s—and for which it was justifiably proud—deteriorated. Beginning in 1944, however, the federal government launched a series of biennial highway acts that provided increasingly more federal aid to the states to compensate for the lack of work during the war and to assist the anticipated postwar building boom. The Federal Aid peaked in June 1956, when President Dwight Eisenhower signed legislation that created the modern interstate highway system, the largest public works project in history.

The period after World War II was a time of change for the highway department as it adjusted to a new series of federal funding bills. The highway commissioners and department executives sought innovative ways to provide matching funds for ever-increasing amounts of Federal Aid. Clearly the old way of enacting debentures wasn't going to provide enough money to match the millions of federal dollars Montana could potentially receive after the war. Compounding the problem was the federal government's establishment of the National System of Interstate Highways in 1944, which mandated higher construction standards and a higher proportion of Federal Aid dollars than on highways not on the system. As the 1940s ended and the 1950s began, the federal government concentrated more of its attention on the interstate highway system—to the detriment of highways not so designated. Because Montana had only three routes on the interstate system, most primary highways in the state suffered. The highway commission wrestled with this problem until Congress devised a new funding system with the passage of the legislation that created the interstate highway system in 1956. Despite the postwar challenges, the highway department spent over $46 million improving and paving 4,754 miles of primary, secondary and urban roadways between 1946 and 1956.

Preparing for Peace in a Time of War, 1942–1945

On December 8, 1941, the United States declared war on Japan after its attack on U.S. naval forces at Pearl Harbor. Three days later, on December 11, Nazi Germany declared war on the United States. With the nation's entry into World War II, the economic depression that had plagued the country since 1930 vanished as the United States' economy and society geared up for what might be a prolonged and costly war. The federal government changed its priority from economic recovery to the war effort. In the process, the Montana Highway Department, which had enjoyed the economic benefits of federal generosity for over a decade, suddenly found itself without federal funds and without a construction program.

The department's ongoing improvements to the state's highways came to a screeching halt. Department employees left to enlist in the armed services or take jobs in the war industries on the West Coast. The highway commissioners encouraged employees to leave the department on the promise that their jobs would be waiting for them when the national emergency ended. By 1943, the number of highway department employees had dropped to 746 people from a high of just under 2,000 people in 1940. In all, 224 highway department employees served in the military during the war. The federal government imposed rationing on rubber, gasoline, oil, steel and other materials critical to the war effort. That made it difficult, if not impossible, for the department to obtain those materials for road construction and maintenance in all but special cases. Travel, gasoline and rubber restrictions for the public contributed to a significant drop in gasoline tax revenues—enough so that the state could not have matched federal funds even if there were funds to match. Because of manpower and equipment shortages, the highway department confined its activities to the limited maintenance of the road system for the duration of the war.

Congress enacted the Defense Highway Act in 1941. It established the National Strategic Network of Highways. Prior to the war, projects on strategic highways had priority for obtaining federal funding and materials to expedite their improvement. Beginning in December 1941, however, the newly created War Production Board and the Office of Defense Transport decreed that no domestic projects, unless approved by the secretary of war, would be authorized except for those on the strategic network or on non–Federal Aid roads. In Montana, the strategic network included U.S. Highways 10 and 91 and U.S. 87 from the Wyoming border to Billings. Only one non–Federal Aid highway was included in the system: the secondary road from Columbus south to the Benbow and Mouat chrome mines near Nye in Stillwater County. The Stillwater Complex in the Beartooth Mountains of south-central Montana was the only known source of chrome ore in North America, a material critical to the war effort. Between 1942 and 1945, the highway commission received only two approvals from the secretary, both for the access road to the chrome mines.

The operations of the Montana Highway Department bottomed out in 1943. The War Production Board suspended construction on all federally funded highway projects in the United States that were not essential to the war effort. The secretary of war approved no projects in Montana. The highway department relied solely on its dwindling gas tax revenues to keep the state's Federal Aid roads in tolerable condition. But despite the highway department's best efforts, the deterioration of the state's roads accelerated during that period. During the winter of 1943–44, the department could afford to plow only the roads to the U.S. Army Air Force's airfields at Lewiston, Cut Bank and Glasgow, along with sections of the primary system deemed important by the War Department. The highway commissioners warned that "persons living in areas that may be temporarily blocked [by snow] should be…cautioned to be prepared for such an eventuality."[37]

The counties couldn't afford the costs of maintaining the secondary road system and looked to the state and federal governments for help. The War

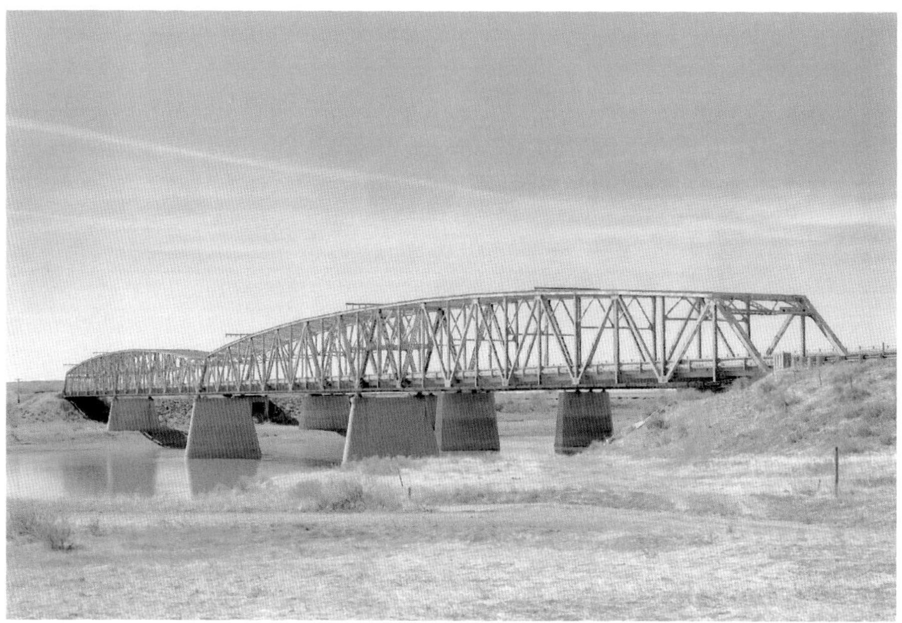

The Yellowstone River Bridge at Fallon was one of the few projects constructed by the Montana Highway Department during World War II. *MDT.*

Production Board allocated materials for highways and bridges based on the priorities established by the National Strategic Highway System. Most primary and secondary roads in Montana wouldn't qualify for construction funds on that basis. Fortunately, however, Congress passed an amendment to the 1941 Federal Highway Act called the Defense Highway Act. The legislation sought to bridge the gap between wartime and postwar programs but also included provisions to reimburse state, county and city governments for road damage caused by war-related activities. The provision provided some relief for the beleaguered state and county governments, but for the most part, maintenance by skeleton crews with worn-out equipment could not keep up with the deterioration. The Defense Highway Act allowed the states to use leftover funds from the 1941 act for engineering and economic investigations and for the preparation of plans, specifications and estimates for postwar construction projects. Importantly, the legislation also authorized a survey to "determine the need for a system of express highways throughout the United States." That provision would be incorporated into federal highway legislation in 1944.[38]

The Federal Aid Highway Act of 1944

Despite the continuation of wartime policy in 1944, there was light at the end of the tunnel. The Allies made significant military gains against the Axis powers in 1943, and it became obvious to most Americans that the war would be won. Even with that optimism, state highway expenditures dropped to a wartime low of just under $4 million (as compared to the high of just over $10 million in 1940). With an eye to the future, Congress passed the Federal Aid Highway Act of 1944, which provided the framework for the postwar highway boom. The act had a profound impact on Montana as it significantly increased federal appropriations for the first three fiscal years after the war and forced the highway commission to seek alternative methods of providing matching funds for the federal money.

The 1944 legislation appropriated $1.5 billion to the states over a three-year period beginning in the first fiscal year after the end of the war. Congress intended the bulk of the money to go to the improvement of the Federal Aid system, with substantial amounts allocated to roads within urban centers and secondary roads. Christened the Federal Aid Secondary System in February 1942, in Montana it comprised 103 routes encompassing 1,833 miles. The highway department selected secondary roads in consultation with the Public Roads Administration and the counties. The states funded improvement projects on the secondary roads, but the counties were responsible for the maintenance of the facilities. The act also made the federal government responsible for one-third of right of way costs and provided 100 percent funding for railroad grade separation projects.

Perhaps most significantly for the future, the act created the National System of Interstate Highways. Although Congress intended the provision to "satisfy the demand for long-distance highways," it provided no money to explicitly develop the system. In June 1941, President Roosevelt created the National Interregional Highway Committee to study and report on the needs for an interstate highway system. Composed of federal and state highway officials and planners, the committee submitted its report, *Interregional Highways*, to the president on January 1, 1944; he forwarded it on to Congress two weeks later. One historian has called the report "the most significant document in the history of highways in the United States" because it marked a turning point in the planning and design of highways to provide, for the first time, a truly national network of interstate highways. Congress incorporated most of the recommendations in the report into the 1944 Federal Aid Highway Act.[39]

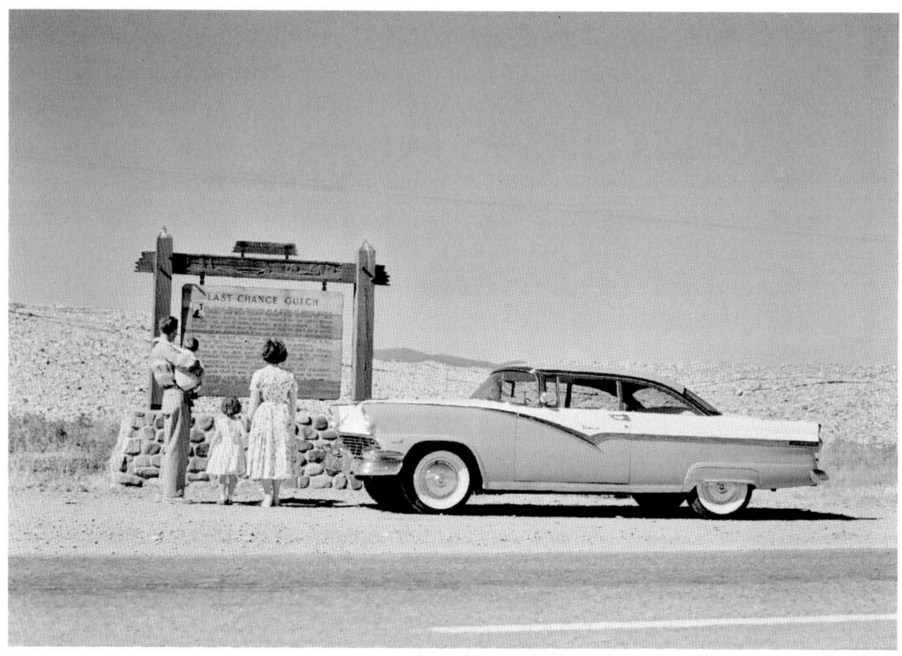

After a four-year hiatus during the war, the highway department revived its tourism programs, including the roadside historical markers. *MHS Photograph Archives, Helena, PAc 76-82.6-A-142.*

The forward-thinking provisions of the Federal Aid Highway Act of 1944 stemmed directly from planning efforts on the part of federal, state and local officials. Not only did it provide for the repair and modernization of the nation's highways after the war and sow the seeds of an interstate highway system, but it also planned for the secondary "farm-to-market" roads. The only problem not addressed directly in the legislation was how the states, including Montana, would be able to raise enough revenue to match the anticipated federal funds. After years of economic depression during the 1930s and reduced gas and license tax revenues during the war, the states were not prepared to provide enormous amounts of money to match federal funds. Montana, at first, responded in its traditional way by enacting a debenture, but that system would be unsustainable during the coming years.

Through extensive lobbying by Governor Sam Ford, the highway commissioners and interested civic organizations, the legislature passed a new $12 million highway debenture bill in early 1945. Ford announced in March 1945 that a special election would be held on June 5 to authorize the highway commission to issue bonds to match Montana's anticipated $46 million in federal appropriations. Beginning in April, commission chairman Al Winkler initiated an intensive media campaign to garner support for the referendum. According to the highway commission, the debenture would not impose any new taxes or result in the raising of the gas tax. Despite the lack of personnel and the nascent highway program during the previous four years, the highway department had not been sitting on its hands. Winkler told the *Helena Independent Record*, "We have on the shelf plans today sufficient to start such a program tomorrow if it were possible to do so. We have sufficient contractors in Montana to execute such a program. These contractors are ready to go at the drop of a hat."[40]

Montanans approved the debenture by a five-to-one margin in June 1945. The proposed postwar program included $23.5 million for construction on the primary system, $3 million for railroad grade crossing elimination projects $1.6 million for construction in cities of more than five thousand people and $17.7 million for secondary roads projects. Indeed, secondary

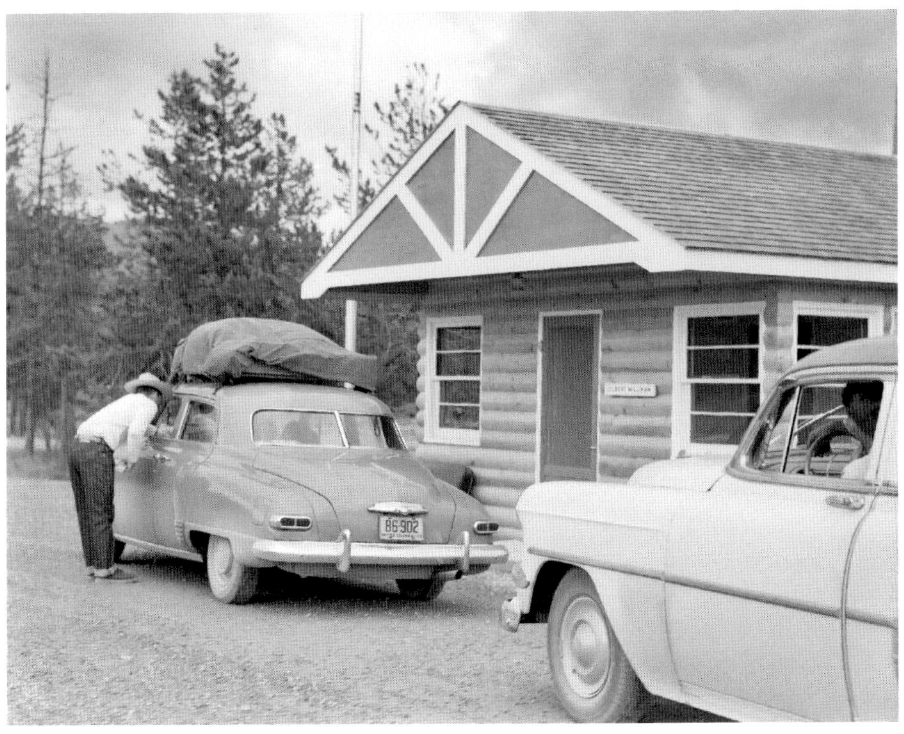

Port of entry stations were important to Montana's roadside landscape until 1958, when the highway department closed them. *MHS Photograph Archives, Helena, PAc 76-82.6-D-402.*

roads were an important part of the postwar program. When the war ended in September 1945, the Montana Highway Department was poised for the greatest road construction program in its history.

From War to Peace, 1946–1956

The postwar years were a time of adjustment for the highway department as it reestablished its prewar programs. While dozens of employees left the department to enlist in the armed forces or work in the war industries, those who remained behind conducted surveys and developed plans for the anticipated postwar program. The Federal Aid Highway Act of 1944 established the foundation for a vastly expanded program for the states. With the end of the war in September 1945, the department had many shelf projects ready to go as soon as federal funds were available. In March 1946, the highway commission had its first big postwar letting when it awarded contracts in the amount of just over $591,000 for six projects. It was truly a period of adjustment, however, as the commission and department dealt with the debilitating effects of a nationwide steel strike in 1946 that significantly hampered the contractors' ability to obtain steel for bridge projects. The Montana Contractors' Association (MCA) and the highway commission also had its share of issues. The highway department's engineers' estimates for projects were still based on wartime prices, while the MCA's members based their estimates on current prices for materials. Consequently, from 1945 to 1948, the highway commissioners rejected over 20 percent of the 205 projects bid on by contractors because their bids were significantly higher than the engineers' estimates.

The department substantially expanded to accommodate its augmented responsibilities. In 1946, the highway commission authorized chief maintenance engineer Ray Percy to purchase over half a million dollars of new maintenance equipment—its first such purchases since 1940. The Maintenance Division also constructed additional shops and offices at Browning, White Sulphur Springs, Glendive, Townsend and Butte. The department established FM radio control stations on MacDonald Pass and in Missoula in 1949 and developed a policy regarding access to drive-in movie theaters. In 1950, the Bureau of Public Roads recomputed Montana's 7 Percent System and added additional routes to the state's road system.

The Cold War also played a significant role in the highway department programs. In 1951, the highway commission entered into an agreement with the United States Air Force to maintain roads and construct prefabricated buildings at four proposed U.S. Air Force radar installations in the state. The department would be responsible for maintaining and plowing the roads and be reimbursed by the air force. The MDT's maintenance men also kept a sharp eye out for Soviet bombers as members of the Ground Observer Corps. Formed in 1950, the corps consisted of civilians trained by the air force to watch the skies and report on any unidentified aircraft they saw or heard. In areas where there were no towns and even fewer residents, the U.S. Air Force relied heavily on the 190 "mobile units of the State Highway Maintenance Department to report aircraft." Unfortunately, the total number of sightings radioed in by maintenance workers is not known, but they did plug a significant gap in the corps' coverage of the state for several years.[41]

Many of the postwar highway projects involved re-routing older highways to improve substandard alignments. Today, Montana's landscape abounds with the physical evidence of that process, specifically on the highways between Three Forks and Augusta and along Montana 200 between Armington Junction and Belt. The highway commission placed new roads on the Federal Aid System, including the segment of U.S. 191 adjacent to Yellowstone National Park in Wyoming. It is the only instance of a highway in another state that is on Montana's Federal Aid system. In 1956, the commission added the secondary road from Billings to Lavina to the primary system to garner extra federal funding for that important trucking route. The 1955 Montana legislature augmented the highway department's programs by empowering the highway commission to designate controlled access highways on the National System of Interstate Highways and allocating $1 million for the construction of costly bridges over the Yellowstone River at Glendive, Miles City and Forsyth and the Clark Fork at Missoula. Throughout the early 1950s, the highway department expanded its infrastructure to include new district office complexes and maintenance shops.

The Secondary Road System challenged the highway department as it added approximately 2,260 miles to the 4,667 miles of primary roads it was already responsible for in 1947. Unlike the primary system, however, the department built the roads, but the counties maintained them—if they had the funds to do so. Secondary road construction peaked between 1948 and 1953, when the highway commission awarded 206 projects totaling 1,428 miles for secondary roads for nearly $24 million, more than half the amount allocated for primary highway construction.

By the onset of the interstate highway era in 1956, the total mileage of Montana's 7 Percent System consisted of 5,926 miles of highways, 94 percent

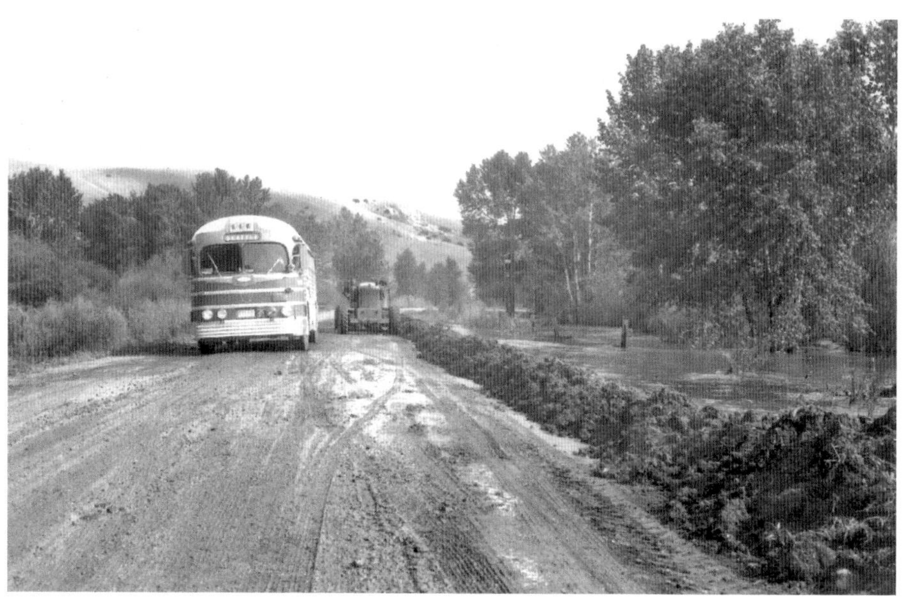

Road construction on U.S. 12 near Garrison, 1948. *MDT.*

of which had received initial improvements of some type. A little over 80 percent had oil surfacing or better, and 9 percent of the roads had gravel surfaces. In the 1930s, highway department engineers designed oil-surfaced roads for a ten-year life, while the substructure was good for fifteen to twenty years. Much of the state's primary system had exceeded that period by 1950. The highway department did its best to reconstruct and maintain the system but found itself falling hopelessly behind. The federal government provided more money to compensate for the deterioration by increasing the biennial appropriations to the state. Montana's tax structure, however, had not changed significantly since 1929, while highway pattern use had substantially changed. Clearly, the department needed new funding sources, but other than increased taxes, there was no clear-cut solution to the problem.

Funding Problems

Finding money to match Montana's share of the five Federal Aid Highway Acts between 1944 and 1956 was the highway commission's overriding concern during that period. While state gasoline and license tax revenues in the 1920s and the debentures during the 1930s proved sufficient to provide the needed funds to match the federal appropriations, it was wholly inadequate to the task beginning in 1950. The enormous amounts of money available to the state for highway construction projects required more money than the funding system could handle. Consequently, for nearly a decade, the highway commission, the governor and concerned Montana citizens wrestled with the problem until they hit on a solution just before the beginning of the interstate highway era—when the funding formula changed once again.

Between 1944 and 1956, federal appropriations to the states for highway construction rose 41 percent from $500 million to $850 million. Congress appropriated the money on a formula that included funds targeted specifically for primary, secondary and urban roads. In 1949, Montana passed its last debenture bill to provide $5.5 million for matching funds over a two-year period. When Congress enacted the 1950 Federal Aid Highway Act, Montana received $7 million. The day of the debenture had passed, and it was time to seek new funding sources to obtain the federal money.

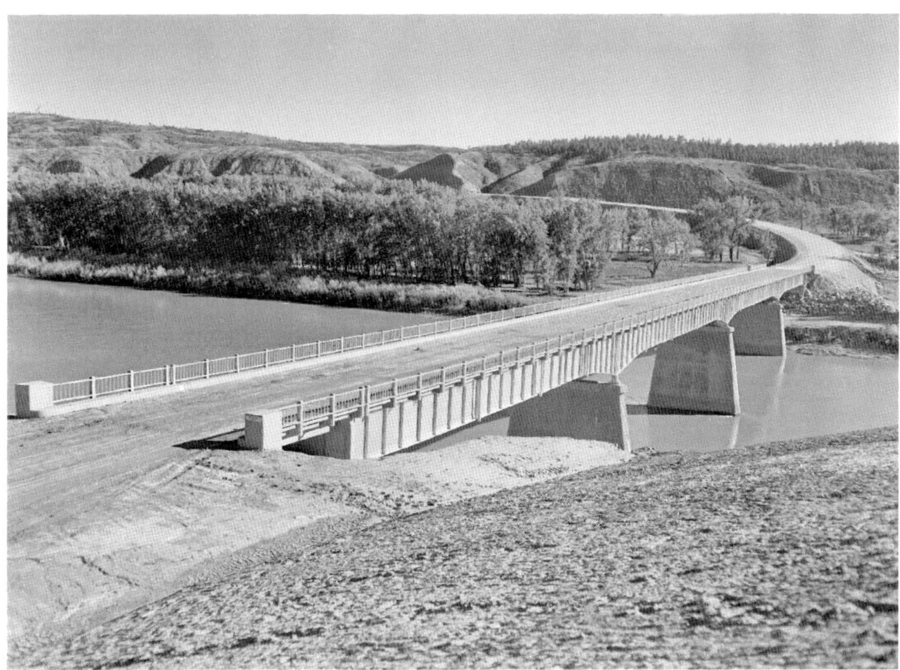

The Fred Robinson Bridge marked the culmination of an effort by central Montana residents for a bridge across the Missouri River. *MHS Photograph Archives, Helena, PAc 86-15.100-B.*

In 1948, Governor John Bonner appointed a statewide committee composed of private citizens to recommend solutions for the highway-funding problem. Over the ten-year history of the committee, the consistent suggestion made by the group was to raise the gasoline tax from six cents a gallon to seven cents a gallon. Along with flat fees for registering automobiles and tractor-trailers, gross vehicle weight taxes, a highway users tax and a petroleum tax, the committee believed the state would have enough money to match federal funds. Legislative reception of these ideas ranged from positive to cold depending on projected revenues. While the committees looked at long-range goals for the state's highway system, the legislature's actions were much more shortsighted. Indeed, until the initiation of the interstate highway era, all "fixes" by the legislature were temporary in nature and failed to come up with any long-term solutions. Fortunately, Federal Aid legislation didn't require the matching money up front, allowing the states grace periods of up to three years to raise the money. Montana, like its sister states, took advantage of the federal government's munificence, but just

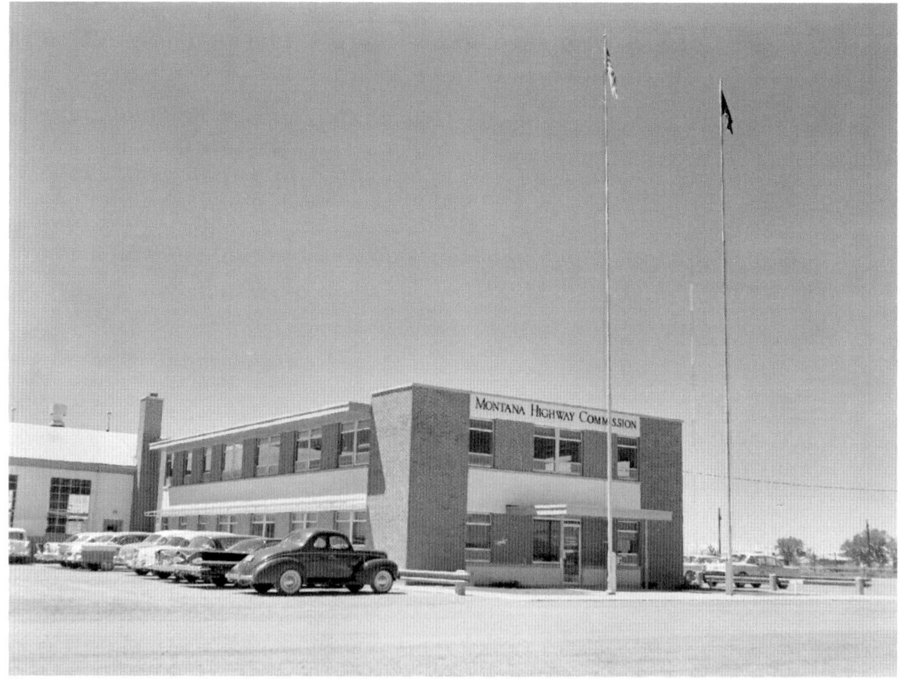

The highway department built offices in its construction districts to handle the increased workload, including one at Billings. *MHS Photograph Archives, Helena, PAc 86-15.96-B.*

barely. The state highway commission's annual expenditure rose from just over $7.5 million in 1946 to $29.3 million in 1956, an average increase of 15 percent per year for ten years—although a $4.7 million increase in Federal Aid in 1956 over 1954 levels meant an increasing demand on the highway commission at the dawn of the interstate highway era.

The National System of Interstate Highways

Serious discussions about a national controlled-access interstate highway system resurfaced just prior to Pearl Harbor. In October 1940, the highway commission began a series of debates about the establishment of a Strategic System of Military Defense Highways. The proposed system, which was implemented by the Public Roads Administration (PRA) in April 1941, prioritized primary roads based on their strategic importance in case the country should become involved in hostilities overseas. Priority One routes received priority in funding and materials for highway and bridge projects, while the Priority Two and Three routes would receive it in descending order of national defense importance. The Priority One roads provided the basis for the interstate highway program of the 1950s, eventually designated as Interstates 15, 90 and 94.

The Federal Aid Highway Act of 1944 included a provision for the designation of the National System of Interstate Highways. Indeed, the system would consume more and more federal and state energy in ensuing national highway legislation, culminating in the Federal Aid Highway Act of 1956. The early years of the interstate highway era consisted of the establishment of the general routes of the interstates, developing standards for projects constructed on the system and the actual construction of the interstate system. Congress, moreover, increased Federal Aid appropriations every two years primarily to compensate for the increasing demands placed on the states by the PRA in regards to the National System of Interstate Highways.

The PRA and AASHO also developed the standards for the new interstate system. In 1947, they established a thirty-two-foot surface width in mountainous areas and a forty-four-foot width on the routes everywhere else. The dispute about interstate roadway widths came to a head in April 1947, when the highway commission refused to design a forty-four-foot-wide roadway on U.S. Highway 10 South in Powell County. It argued that traffic in Montana, "with two or three exceptions contiguous to the

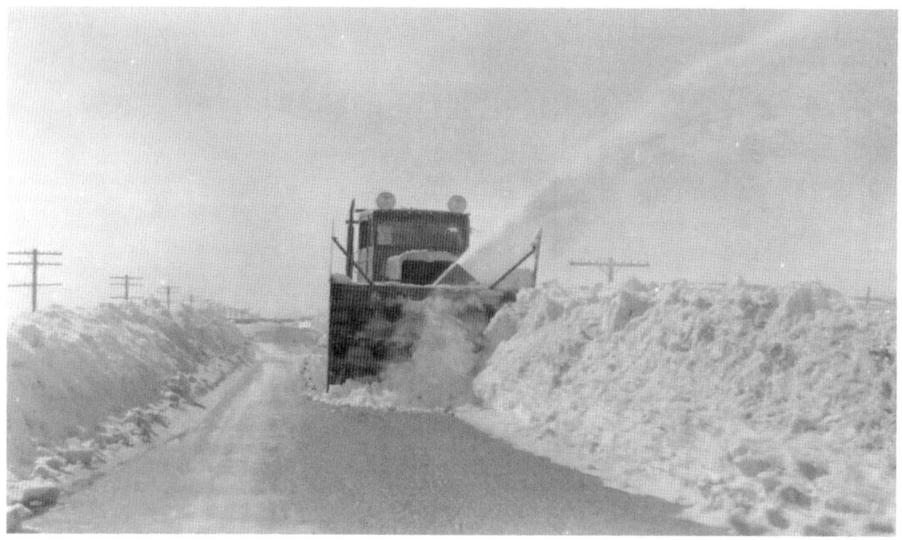

The Montana Highway Department initiated a statewide snow-plowing program after World War II. *MDT.*

population centers, was so low that the standard proposed by the BPR could not be justified on any basis, either of traffic needs or of economy." The commissioners suggested that if the federal government were willing to pay the difference, the highway commission would construct a forty-four-foot-wide paved highway. Eventually, however, the PRA prevailed, and the Butte-based F&S Contracting Company constructed a forty-four-foot-wide highway during the summer of 1948.[42]

Generally, the highway commission supported the National System of Interstate Highways but disputed the standards imposed on it by the federal government, which did not take into account Montana's sparse population and the fact that the state had the "third longest mileage of Interstate highways in the United States." Beginning in 1952, the Federal Aid Highway Acts allocated funds specifically for the interstate system, which had to be matched by the states in the regular sixty-forty ratio. The legislation stipulated how the money could be spent on the system and for what purposes. The highway commissioners had ceased arguing about the design standards by 1952 and accepted the development of the system as part of the regular Federal Aid. In February 1956, just prior to the passage of the landmark Federal Aid Highway Act legislation formally creating the Interstate Highway System, the highway commission designated Montana's first control-access highway, U.S. Highway 87 in Big Horn County.[43]

Montana's White Crosses: The Roadside Fatality Markers

In September 1952, Louis Babb of American Legion Hellgate Post #27 in Missoula appeared before the Montana State Highway Commission and proposed a project that would memorialize traffic deaths on Montana's highways by erecting white crosses at fatal accident sites. Through noting the sites of fatal accidents, the American Legion hoped to "call the attention of the public to the necessity for better and safer driving." The commission agreed to the plan as long as the crosses didn't interfere with snow plowing or maintenance activities. Six months later, in April 1953, Babb and Herb Kibler reaffirmed the commission's approval, formally initiating a program that has been active in Montana for over sixty years.[44]

Floyd Earhart conceived the idea of the white crosses to mark the site where six lives were lost in Missoula County during the Labor Day

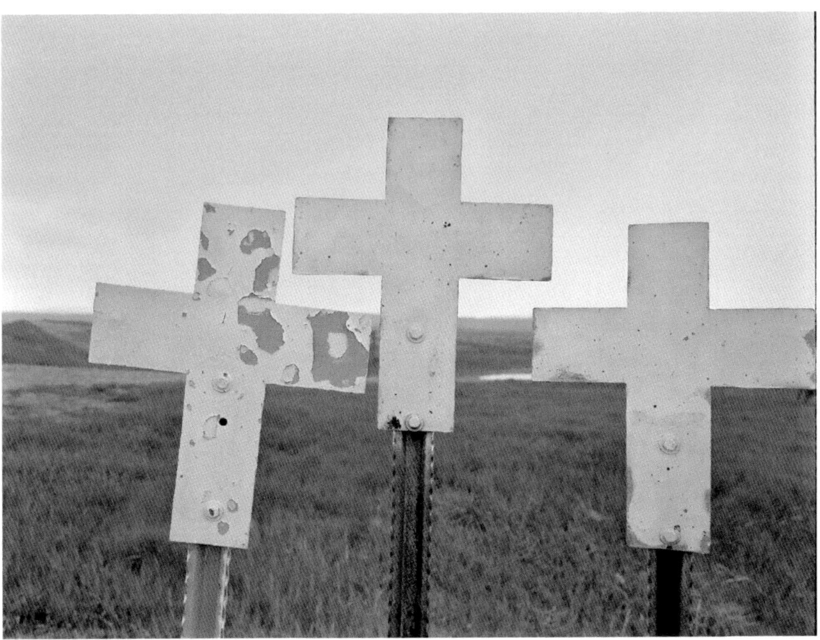

Roadside white crosses have been a stark reminder of Montana's highway fatalities since 1953. *Lucy Capehart.*

holiday in 1952. It initially began as a county project but had quickly expanded statewide by 1953. Along with the endorsement of the highway commission, every Montana governor, beginning with J. Hugo Aronson, has sanctioned the project. Most of the state's American Legion posts participate in the program and are responsible for the installation and maintenance of the crosses. Contrary to popular belief, they do not mark graves, but the more than two thousand markers erected by the Legion would fill a five-acre cemetery. White crosses are permitted within the rights of way of primary, secondary, urban and U.S. Forest Service roads; they are not permitted within interstate highway rights of way. The program is not intended as a memorial so much as a "sobering reminder of a fatal traffic accident, a place where a human being lost his/her life."[45]

For the ten-year period between 1946 and 1956, the highway commission reestablished its prewar programs, devised new funding formulas to match Federal Aid and geared up for the Interstate Highway program that would dominate the highway department for the next three decades. In some respects, the highway programs of the Great Depression proved a transitory period for the postwar boom in road construction. From 1956 onward, the commission concentrated on design and construction of the interstates.

PART VI

YOUR HIGHWAY DOLLARS AT WORK

THE INTERSTATE HIGHWAYS

It hasn't been easy to overcome all the difficulties posed by the Interstates, and only through a concerted effort by all divisions, has it been possible to reach the level we are on now, and should be able to remain on till it is completed. Every division should take pride in the output of the fellows who have contributed to the effort.
—*Fred Quinnell Jr., 1961*

Few things have had as significant an impact on the United States and Montana as the interstate highway system. Arguably the largest public works project in history, the interstate highway system encompasses over forty-seven thousand miles of highways in the United States and cost American taxpayers an estimated $425 billion to build, with annual maintenance costs exceeding $50 million. Originally envisioned by the federal government as a means to expedite commercial, military and civilian traffic, its construction was a monumental task for the federal government and state transportation departments. During the planning, design and construction phases of interstate projects, federal and state officials contended with routing and environmental issues, the acquisition of enormous amounts of right of way, a public not always supportive of the projects and political opposition. Construction presented challenges in placing four-lane highways in areas not always suitable to accommodate them. Building the interstates was often as much about selling the highways to the public as it was the development of innovative design and construction techniques.

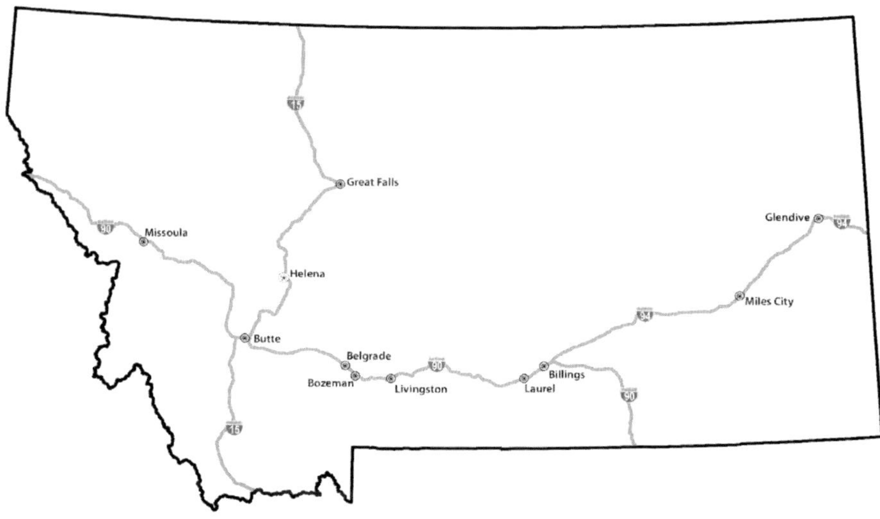

Montana's interstates.

Although an interstate highway system had been the goal of the federal government since at least the 1920s, actual planning for it didn't begin until just before World War II. In response to a request from Congress in 1939, the Public Roads Administration (PRA) and American Association of State Highway Officials (AASHO) published a report, *Toll Roads and Free Roads*, which documented the need for a system of interregional superhighways with connections through and around American cities. A follow-up study in 1944, *Interregional Highways*, conveyed the need for a nationwide system of controlled-access highways encompassing some 39,000 miles (including 33,900 miles in the country's rural areas). Congress incorporated both documents into the Federal Aid Highway Act of 1944, which designated the National System of Interstate Highways. The system, based on the National Military Strategic Network, delineated a 40,000-mile national highway system that connected metropolitan and industrial centers. By 1947, the PRA, AASHO and the states had mapped out the system; in Montana, this matches today's interstate system.

After years of intense lobbying and revisions, President Dwight Eisenhower signed the Federal Aid Highway Act of 1956 into law on June 29, 1956. The legislation created the National System of Interstate and Defense Highways and specified how the system would be constructed and financed.

It stipulated a uniformly designed interstate highway system for the first time in U.S. history. The act allocated $25 billion to the states over a thirteen-year period to build the system. It required controlled-access, twelve-foot travel lanes and ten-foot paved shoulders; the elimination of at-grade railroad and road crossings; separated traffic lanes; and seventy-mile-per-hour speed on flat terrain.

Building the interstates was, indeed, a challenge but one the Montana State Highway Commission and Highway Department eagerly accepted—a once-in-a-lifetime opportunity for the department's engineers. Interstate construction occurred in two phases. Initially, the engineers designed the interstates as two-lane facilities, except near metropolitan areas where highway traffic counts warranted four-lane highways. Between 1958, when the highway commission let the first interstate contract, and 1966, when the Federal Aid Highway Act mandated that all interstates be four lanes, the highway department built 491 miles of interstates. Then it had to go back and expand the two-lane segments into four-lane segments.

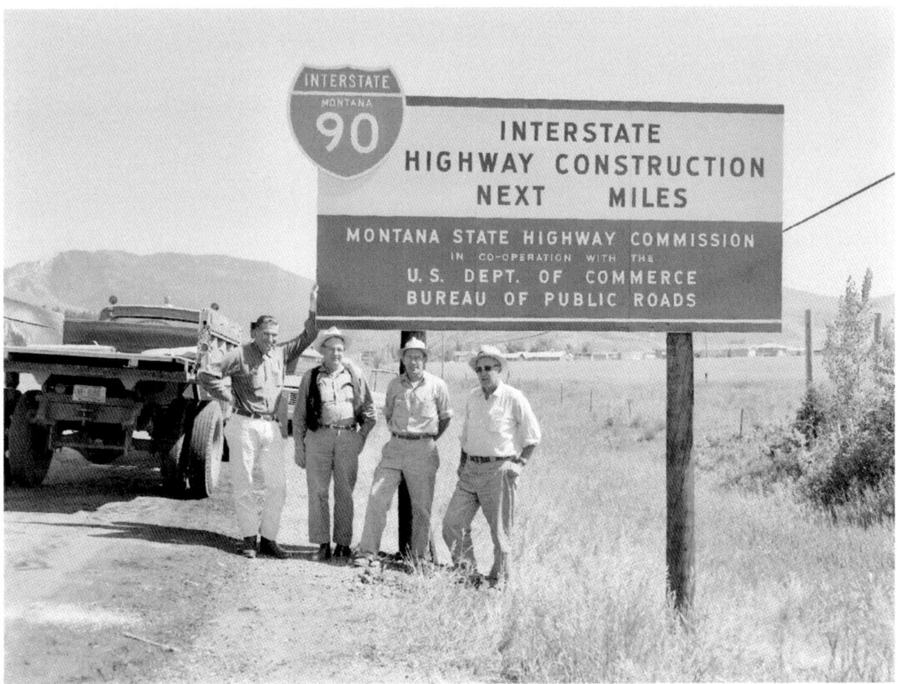

Interstate construction signs were a common sight in Montana from 1959 to 1988, when the last segment of highway was completed. *MHS Photograph Archives, Helena, PAc 86-15.6182-A.*

The interstate highways in Montana had a profound impact on the state. The planning, preconstruction and construction phases caused the greatest expansion of the Montana Highway Department in its history. Montana contractors benefited mightily from the federal funds used to build the highways. The highway department's snow removal and maintenance practices expanded to take care of the system. The interstates caused a significant change in how the state did business as the emphasis shifted away from railroad freight traffic to commercial trucking. Impacts on the environment by the interstates resulted in the enactment of stricter laws to protect wetlands, fisheries and water and air quality. The federal government also legislated protection for parks and historic and archaeological sites threatened by the interstates. Like the railroads in the nineteenth century, the interstate benefitted some communities at the expense of others bypassed by the highways. Still other towns shifted their socioeconomic impetus from local trade to function more as bedroom communities for larger metropolitan areas. There were also changes in the landscapes of the cities as commercial development migrated to the new interchanges.

The Golden Age of Montana's Interstate System, 1956–1966

The general route of Montana's interstate system had been established for nine years by the time President Eisenhower signed the Federal Aid Highway Act into law in June 1956. While the routes had long been set, the details had not. There were still questions about alignments, two-lane versus four-lane facilities, problems with the concept of controlled access and, most of all, how to fund Montana's share of the construction of nearly two thousand miles of superhighways. A sparsely settled state, Montana was fortunate in its large amount of federally owned lands as compared to other states. Because of that, the Bureau of Public Roads (BPR) designated Montana a public lands state, which meant that the federal share of its allocation was based on the amount of federally owned or administered land within its borders. For Montana, the ratio was 91:9 percent. Even so, the prospect of raising the needed matching funds was a formidable one for the Montana state legislature and the state highway commission.

To carry on the interstate program, the highway department beefed up its Road Design Section with additional engineers and designers. Former highway department engineer Steve Kologi remembered there was an effort to recruit engineers from the field. The department "took the cream of the crop…of the project engineers, brought them to Helena and made designers out of them." The new designers had plenty of experience in highway construction, making them valuable assets to the program. In 1958, the department completed construction of a $600,000 addition to the highway department building on the state capitol complex to accommodate the expanded workforce.[46]

The highway department's Interstate Division planned for the interstates, particularly their alignments, and conducted public hearings. The department's Right of Way Bureau expanded to handle the enormous task of acquiring rights of way for the interstates. Directed by Lewis Chittim, his tenure had a significant impact on the design and construction of Montana's interstate system. In April 1957, the highway commission appointed Fred Quinnell Jr. as the new state highway engineer, replacing longtime department engineer Scott Hart.

At the dawn of the interstate highway era, Montana had 1,194 miles of interstate highway divided into five routes: Interstates 15, 90, 94, 115 and 315. Initially, highway planners determined that four-lane sections of interstate would be constructed only where the Design Hourly Volume (DHV) was greater than seven hundred. That meant four-lane sections would be located only at Missoula, Butte, Bozeman, Billings, Miles City, Helena and Great Falls, with the bulk of the state's interstates consisting of two-lane, controlled-access routes in the state's rural areas. Highway department engineers estimated that Montana would require near half a billion dollars to complete 1,194 miles of interstate. That included 1,178 miles of two-lane interstate and 16 miles of four-lane highways. Of that total cost, approximately 10 percent would

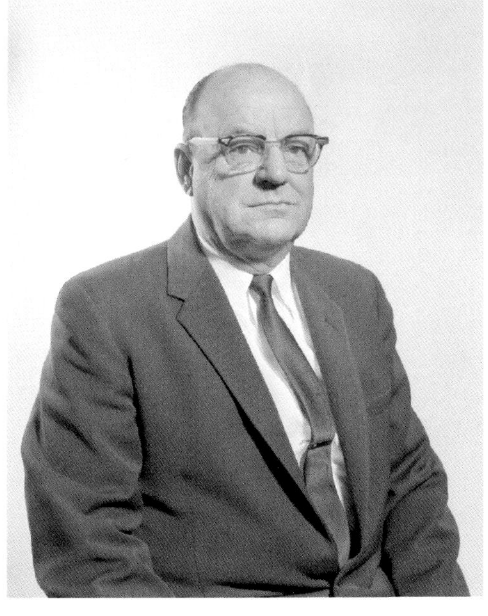

Fred Quinnell kept Montana's interstate highway program on track from 1957 to 1972. *MHS Photograph Archives, Helena, PAc 86-15.51-B.*

be needed for preconstruction and right of way acquisition, with the remaining money targeted for construction. Ultimately, Montana's interstate system would cost $1.22 billion, of which the state provided $111 million to match federal funds.

The gas tax raised a little over $17 million for the state in 1956 and again in 1957. The Montana Highway Department anticipated that the gasoline tax would provide only $12.5 million in 1958. The expected federal appropriation that year would be $40.7 million, of which Montana would need $16.0 million and another $14.0 million in 1959 to match it. Montana was in danger of losing federal funds for the interstate program. An economic recession in 1957, moreover, carried over into 1958. In an effort to stimulate employment, Congress allocated another $400.0 million to the states for road construction. Montana's share of those funds was $36.0 million. Fortunately, the legislature added a penny back on to the gasoline tax in the early 1960s, which helped Montana obtain the maximum federal funds available to the state for interstate highway construction. By 1960, however, Montana again had a backlog of federal funds it was unable to match. That year, Congress provided a new allocation formula that allowed states like Montana to get back in line with the original allocation system.

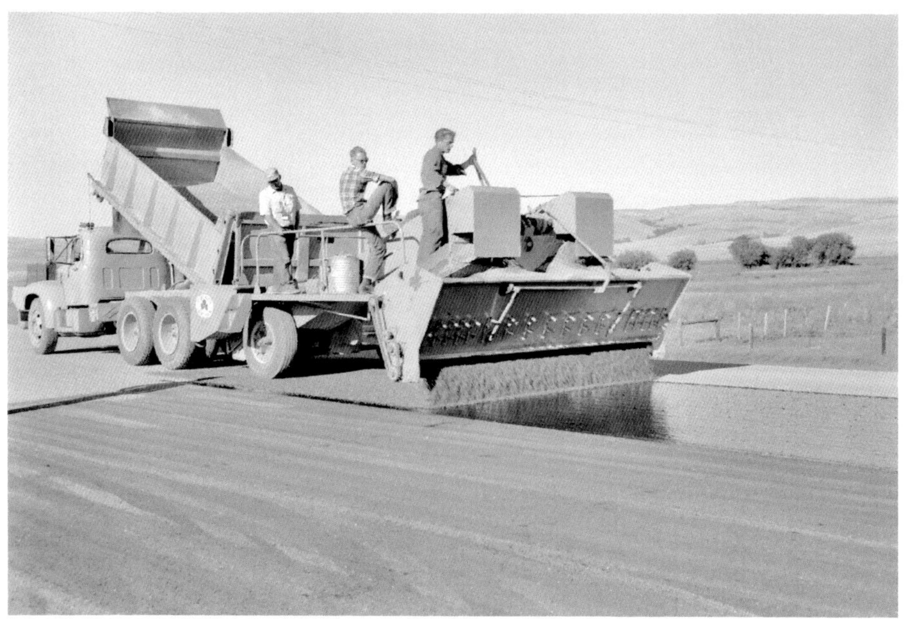

Chip sealing a recently paved segment of I-90 near Deer Lodge in 1962. *MHS Photograph Archives, Helena, PAc 86-15.61222-F.*

Despite the funding issues, planning for the interstate system in Montana rapidly progressed. The department's Interstate Division established routes for the new highways. The 1956 Federal Aid Highway Act required that the state highway departments conduct public hearings for each interstate project. The meetings often proved contentious. Many Montanans didn't understand what the interstates really were and had a difficult time with the idea of controlled access. Because the majority of the interstates would be on new alignments, few rural landowners were thrilled when they discovered where they would be located. Battles over alignments were common, which often led to the frustration of the department's engineers tasked with the monumental projects and the public, who didn't want them in their backyards. Steve Kologi later remembered, "There wasn't much Interstate that seemed very easy and Montana had 1200 miles of it."[47]

Nationally, three out of every four miles of interstate were built on new alignments. The designers straightened out and shortened the interstate system as compared to the original primary highways, which tended to fit more within the existing landscape. In order to meet the standards set for it, designers frequently had to make the landscape fit the interstate. While much of the interstate generally followed the old highway system, many planners and engineers saw the interstates as an opportunity to improve alignments and shorten traveling time for motorists and truck drivers.

Some proposed interstate routing problems were not easily solved. In 1957, the engineers recommended that the interstate between Billings and Livingston closely parallel U.S. Highway 10 through the Yellowstone Valley. Residents in that area heatedly protested the decision because it would impact too much farmland. The highway commission reaffirmed its decision in 1958. Nearly a year later, however, a special legislative committee conducted a hearing on the commission's decision. The highway department was represented by state highway engineer Fred Quinnell and Interstate Division chief Grover Powers. The department's newsletter, the *Center Line*, later described the hearing as more of a trial. The committee members didn't feel the interstate should be routed through the Yellowstone Valley but should, instead, follow an alignment farther to the north. The committee asked "leading and misleading" questions but didn't allow Quinnell or Powers a chance to explain their answers. All witnesses produced by the committee were against the Yellowstone Valley alignment. Although the BPR and highway department had surveyed a "northern" route, they found

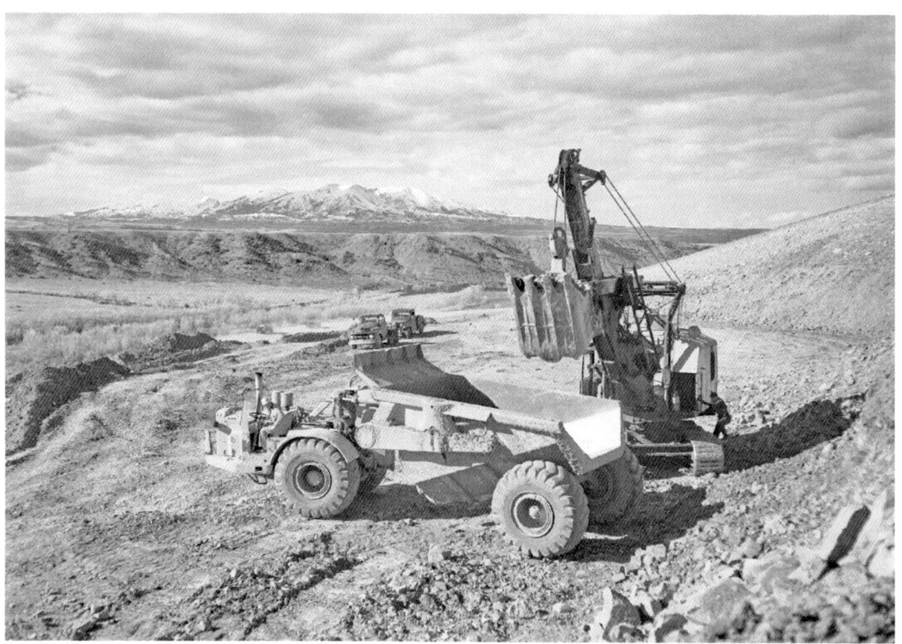

The repercussions of the massive I-90 Springdale Cut put Fred Quinnell in the hot seat during the 1959 legislative session. *MHS Photograph Archives, Helena, PAc 86-15.80-C.*

it impractical. Despite the hostility against the Yellowstone Valley route, the highway commissioners once again restated their support of the route in July 1959, and construction began on it in 1960.[48]

Many decisions made by the highway commission were unpopular with the public. In 1963, the tension between the commissioners and the public peaked at the state legislature in regards to the interstate construction program. In October 1961, the highway commission let the first of six contracts to construct the 10.0-mile segment of I-15 through Wolf Creek Canyon (formerly known as Prickly Pear Canyon) north of Helena. Initially, the highway department engineers investigated several alternate routes, one of which would have avoided the canyon. Ultimately, however, they determined that the "middle" route through the canyon was the most feasible from an economic and maintenance standpoint. As the highway commissioners prepared to let the contract for that segment in January 1963, state senator Ben Stein and six other senators introduced a resolution in the state senate asking that the commission delay further contract lettings on I-15 in Wolf Creek Canyon "pending adequate public hearings and complete engineering review of present routings." The resolution asked that a study be completed to determine whether the preferred routing should be dropped in favor of the east alternative—this after the first 3.4-mile segment of the interstate at the south end of the canyon had already been built at a cost over $4 million. The resolution maintained that the east route would shorten the distance between Helena and Great Falls and involve lower right of way and construction costs. The preferred route, the senators claimed, had been chosen by the highway commission without adequate public involvement and would "drastically curtail fishing and picnicking in the canyon."[49]

The resolution stunned the highway commissioners, Quinnell and several members of the senate's highway committee. Not only did it jeopardize all the projects in the canyon, but it also suggested that the already built segment of I-15 from Sieben to the south end of the canyon be abandoned for a totally different alignment. Quinnell believed that the state could lose

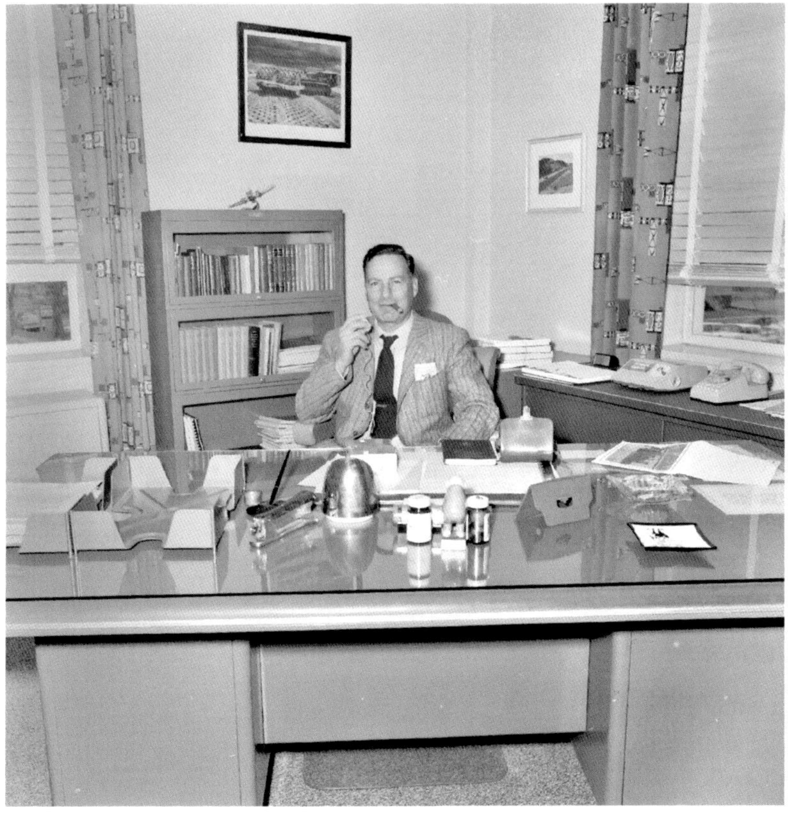

State senator Ben Stein, seen here at Fred Quinnell's desk, opposed the highway department's I-15 plans in Wolf Creek Canyon. *MHS Photograph Archives, Helena, PAc 86-15.6333-B.*

$10 to $12 million if that routing were dropped in favor of a new alignment. During Quinnell's testimony before the committee, Stein challenged the highway commission on the issue of the public hearings, stating, "To me these hearings are a fake. They are the closest thing to an iron curtain we have in America…They [the commission] go on, listening kindly to people with the full knowledge they will do what they want." He claimed the preferred route would destroy "irreplaceable scenic values which are unexcelled any place in the northwest outside of the National Parks."[50]

Quinnell vehemently disagreed with Stein's assessment, asserting that the project actually enhanced the scenic qualities of the canyon. The chief engineer wasn't the only one annoyed by the committee's resolution and comments. State senator Charles Bovey of Great Falls, chairman of the highway committee, didn't support the resolution. Instead, he urged the commissioners to let the next segment of interstate to contract before the resolution passed the legislature. He claimed that it was important for the state's economy that the project proceed as scheduled. The commission heeded Bovey's request and let the $2.3 million project as planned.

The contract letting didn't end the controversy. In a senate highway committee meeting held in February, Senator Stein called Quinnell the "king pin of a department that has been unresponsive to the public which it is supposed to serve" and demanded his resignation. Stein was upset about the way the highway department conducted public hearings in regards to interstate routings. He believed that the decisions had already been made by the time the hearings were held. Stein pulled no punches in regards to the department, calling its employees "a bunch of pettifogging paper shufflers" who wouldn't give the public "the time of day…unless they know where they are on this involved route which construction plans must follow." Senator Bovey took exception to Stein's remarks at the hearing, stating that he believed the way the highway department conducted public hearings was the fault of the legislature, not the department.[51]

Much of the expense for the new highways was tied up in right of way (ROW) costs. The highway department's ROW Bureau expanded in the wake of the 1956 Federal Aid Highway Act. The expansion included additional staff for real estate appraisals and negotiations, but the old system of "horse trading" for ROW no longer worked. Consequently, in 1959, the highway department hired Lewis Chittim as its ROW engineer. Chittim established the state's first appraisal system for the acquisition of ROW. He directed that the department obtain a quitclaim deed for all the land the department acquired as part of its construction program. Despite Chittim's best efforts,

A Tough Piece of Work: Interstate 15 and the Wolf Creek Canyon

Despite the political battles in the legislature, the early 1960s was a golden age for the interstate program in Montana. The Montana Highway Department had considerable latitude in how it designed and built interstate projects. That coupled with what seemed, to many, to be unlimited federal funds created a climate of optimism in the department that was unmatched at any other time during its history. Between 1958 and 1966, the highway commission let contracts for two hundred interstate projects totaling over $24 million. The projects not only changed how motorists drove but also had a substantial impact on the Montana landscape.

During the '60s, the highway commission tackled big projects through the most difficult terrain in western Montana. This included sections of I-90 in Mineral and Granite Counties and between Whitehall and Butte over Homestake Pass east of Butte. The interstate project through Wolf Creek Canyon was the first in Montana to represent all the funding, financial, design, environmental and political issues associated with the program both nationally and locally. It had it all, including routing and construction issues, environmental impacts, combative political opponents and, importantly, substantial impact on a small community unfortunate enough to be in the way of the superhighway. The winding route through the narrow canyon had been a challenge for road builders since Montana's gold rush days a century before and would be again in the 1960s.

Planning for I-15 north of Helena began in the mid-1950s, shortly after the passage of the Federal Aid Highway Act of 1956. The highway department's engineers had established the general route of the interstate between Helena and Great Falls by 1957. Initially, the engineers investigated eleven alternate routes from Helena north to Hardy Creek and eventually cut that number down to three. The Wolf Creek and Missouri River Canyons north of the Capitol City were unique challenges in how best to site the highway while minimizing impact on the water courses, fisheries and the community of Wolf Creek at the mouth of the canyon. Two of the three alternative alignments traversed the canyon, while one route passed close to the Gates of the Mountains, negotiating two narrow canyons and paralleling the east shore of Holter Lake before rejoining U.S. 91 about two miles northeast of Wolf Creek. Ultimately, the highway commission chose the so-called middle alignment through the canyon, the route the interstate follows today.

The highway commissioners chose the middle route because of construction and future maintenance costs, but the opportunity to maintain the old U.S. 91 alignment through the canyon as a recreational access to Little Prickly Pear Creek was also a deciding factor. Although the Holter Lake alignment was two miles shorter than the preferred alternative, it would cost over $8 million more to construct, with maintenance costs significantly higher. Two of the proposed interstate alignments would have an impact on the community of Wolf Creek at the mouth of the canyon. Established in 1887, Wolf Creek historically served as a station on the Great Northern Railway's Montana Central Railroad. But by the late 1950s, much of the town's economic livelihood was based on the recreational opportunities afforded by nearby fisheries.

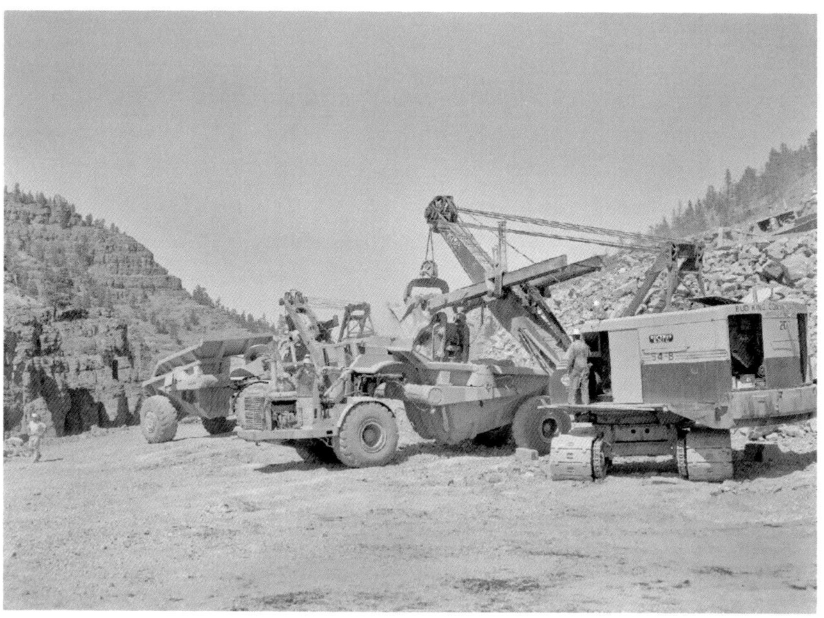

Removing rock after blasting in Wolf Creek Canyon. *MHS Photograph Archives, Helena, PAc 86-15.72177-B.*

Because of the narrowness of the canyon at Wolf Creek, impacting the community with the construction of the interstate was unavoidable. Alternatives at Wolf Creek would destroy either the town's residential district or the businesses that catered to recreationalists. Ultimately, the engineers chose the alignment that would wipe out the town's main street commercial district. They left homes and businesses on either side of the planned route intact. With the routing selected, the highway department held its first public hearing regarding the project.

Highway department engineers had already selected an interstate alignment when it held the public hearing at the Wolf Creek School in March 1961. Speaking to a packed house, the department's engineers described the route of the interstate and its impact on Wolf Creek. Despite the potential for controversy, there was very little discussion about the routing of the interstate through the canyon and Wolf Creek. Most of the discussion concerned the effects on the Little Prickly Pear Creek fishery.

Working in collaboration with the Montana Fish and Game Commission, interstate engineers designed a series of "lifts and drops" in the creek to preserve the trout habitat. The Fish and Game Commission was concerned about what impact straightening the creek to accommodate the highway would have on fish. Local sportsmen's groups echoed the agency's trepidation. Representatives of the Cascade County Wildlife Association and the Central Montana Sportsmen's Association believed the highway would destroy a prime fish habitat. Although the Fish and Game Commission had already reached an agreement with the highway department regarding the creek, it still favored the Holter Lake route. Indeed, the sportsman groups also favored the Holter Lake alternative but conceded that the routing issue had already been settled. Highway department planning survey manager Howard Buswell settled the argument when he stated that it was a "matter of whether the canyon is more important for highways or for fishing."[52]

Beginning in October 1961 and over the course of the next three years, the highway commission awarded six contracts for the construction of the fourteen miles of four-lane interstate and twelve bridges through the canyon. Although traffic counts warranted a two-lane interstate, the highway commission opted to construct a four-lane facility through the canyon. The engineers reasoned that it was more cost-effective to build the four-lane road then rather than build a two-lane facility and go back sometime in the future to widen the corridor to accommodate four lanes.

Construction of the 3.4 miles of interstate through the southern end of the canyon was an enormous and complicated undertaking as the head of the canyon was barely wide enough to accommodate the creek, Highway 91 and the railroad. The highway commissioners awarded the $2 million contract to two companies, the Bud King Construction Company of Frenchtown and McLaughlin, Inc., of Great Falls, to construct the interstate in late October 1961; they began work on the project a few weeks later. Because the narrow canyon was also the route of U.S. 91, the contractors had to keep the old highway open during construction of the interstate. The project required considerable blasting of the rocks at the mouth of the canyon. The contractors posted guards at each end of the construction zone to prevent traffic from passing through when the blasting occurred. Delays ranged anywhere from thirty minutes to over a day depending on the amount of material that needed to be removed after the blasting occurred. While the *Helena Independent Record* and the *Great Falls Tribune* published warnings about the blasting and suggested that motorists find alternate routes, more than a few travelers waited out the delays and occasionally complained to the newspapers about them.

The job of cutting a four-lane road at the south entrance of the canyon was monumental and involved round-the-clock construction. It required the construction of enormous fill slopes and sweeping curves to maintain the fifty-mile-per-hour design speed of the interstate through the canyon. In all, the contractors removed one and a half million cubic yards of material to make way for the highway. When completed, one man called the highway the "seventh wonder of the engineering world because the back slopes at the mouth of the canyon were so extensive." The contractors completed work on this 3.5-mile section of I-15 in late 1962. The highway commission then prepared to let the next contracts, for 11.4 miles of interstate and six substantial bridges, in January 1963.[53]

It was only after considerable political wrangling during the 1963 legislative session that the highway commission let the next segment

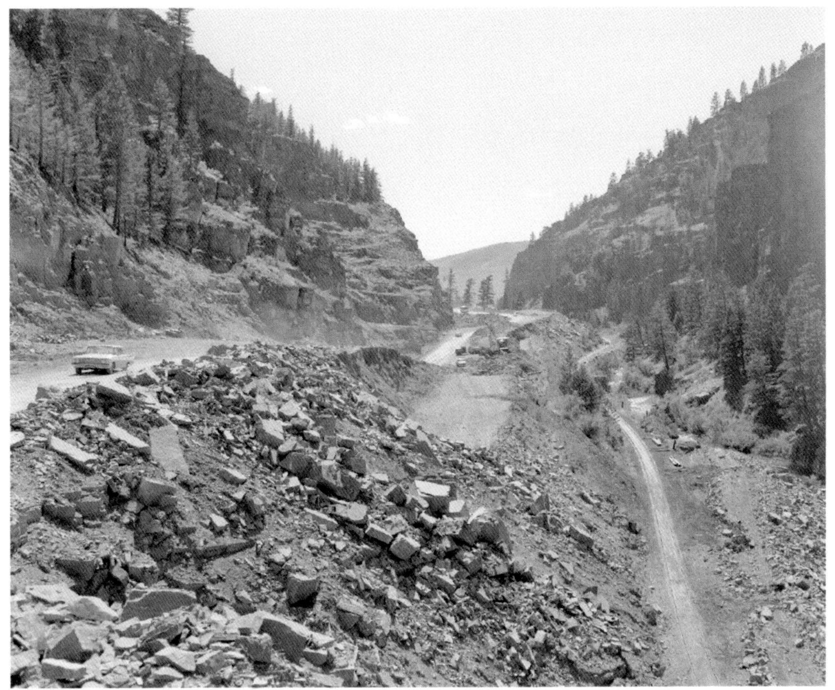

Contractors on the Wolf Creek Canyon section of I-15 kept the old highway open during construction. *MHS Photograph Archives, Helena, PAc 86-15.62177-C.*

of I-15—the $2.3 million Lyon Creek South project—to contract in late January. Despite the issues debated at the legislature, work on I-15 through the Wolf Creek and Missouri River Canyons continued unabated until Helena and Great Falls were connected by a four-lane interstate in 1968. Montana newspapers and the highway department's monthly newsletter, the *Center Line*, kept readers updated on the progress of the projects. The failure of a large culvert on the Lyons Creek South section of the canyon in July 1964 generated newspaper headlines around the state and forced the highway engineers to redesign other culverts on the interstate system. In 1966, the highway department opened its first interstate rest area at Lyons Creek in the canyon. In all, over $10 million was spent on the fourteen-mile Wolf Creek Canyon segment of Interstate 15.

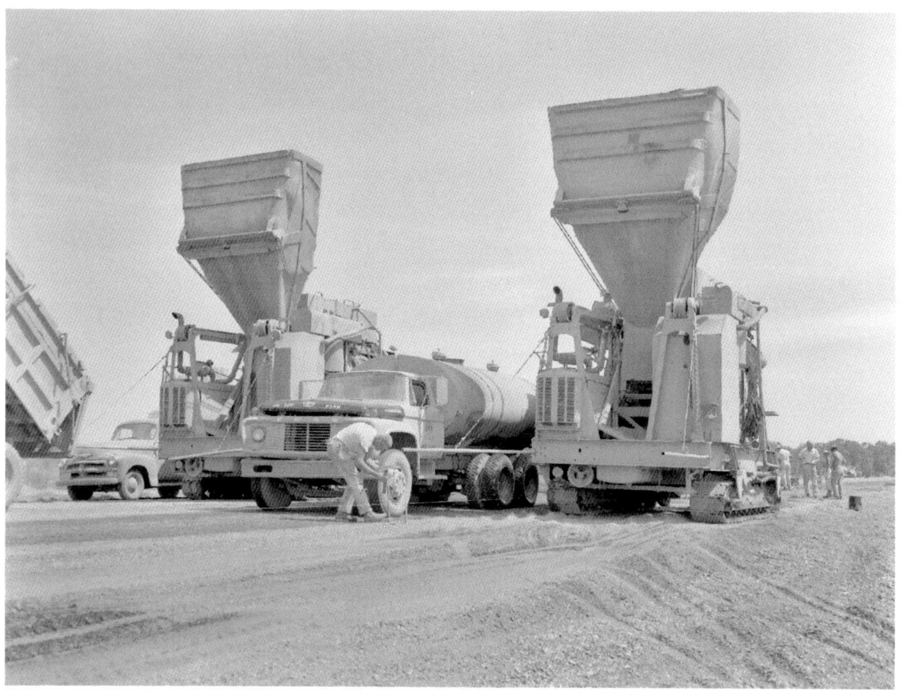

Interstate highway surfacing near metropolitan areas was usually concrete. Pictured is a mobile concrete batch plant in 1961. *MHS Photograph Archives, Helena, PAc 86-15.61091-M.*

ROW issues chronically slowed the construction of new interstate highways in Montana. Fully 10 percent of the ROW needed for the highways was gained through litigation and condemnation.

On March 28, 1958, the state highway commission awarded contracts for the first two interstate projects in Montana. They involved the building of about five miles of two-lane, controlled-access Interstate 15 in the vicinity of Lima in Beaverhead County and the construction of two overpasses. Over the next nine months, the commission awarded eleven more interstate highway contracts. Most of the projects were on Interstate 90 west of Missoula, between Bearmouth and Drummond, outside Billings and between Hardin and the Wyoming border. Although engineers' estimates of how many miles of interstate were constructed in those first years vary, the best evidence suggests that the highway commission awarded contracts for $10,120,146 comprising a little over 43 miles of highway. Over the next seven years, the highway commission awarded over $187.6 million in interstate projects that resulted in the completion of 491 miles of highways.

Interstate engineers designed the highways to accommodate projected traffic demands in 1975. By the mid-1960s, new federal guidelines stated that the highways had to be designed to meet traffic demands anticipated twenty years from the planned construction date. The BPR directed that the highways all have twelve-foot driving lanes, ten-foot shoulders and, where feasible, thirty-six-foot depressed center medians. All interchanges, no matter whether cloverleaf or diamond pattern, had to meet strict geometric standards, and all overpasses had to accommodate fourteen-foot overhead clearances. There were even standards for the height of vegetation adjacent to the roadways. Design speeds varied between fifty miles per hour in mountainous terrain to seventy miles per hour on flat terrain. The BPR allowed no advertising or businesses within interstate ROW.

American interstate designs and standards originated with the German autobahn system before World War II. While Adolf Hitler had intended the autobahns to expedite the movement of troops, military equipment and supplies between the country's eastern and western borders, that was not the case with the American interstates. Bureau of Public Roads chief Frank Turner stated that the interstates were designed to move goods rather than people. Despite that, there has been a persistent myth in the United States that the interstates were designed to function primarily in a military capacity. During the early days of the program, the BPR and the U.S. Air Force considered incorporating long "straightaways" every five miles into the design of the highways for use as runways for warplanes during national emergencies. Although the BPR and air force considered the idea, the BPR soon discarded it as unfeasible. The BPR and the military discovered that the fourteen-foot overpass clearances were not high enough to accommodate the movement of missiles in the western United States. Consequently, the BPR increased the standard overpass clearances to sixteen feet. One aspect of the interstates indirectly associated with the military is their intended use as escape routes from the cities in the event of nuclear war.

In March 1963, the highway commission and Quinnell began discussions with the BPR about the expansion of Montana's two-lane interstates to four lanes. The state highway engineer believed that the "design and construction of short sections of two-lane opposing traffic roads between larger sections of four-lane divided highway certainly cannot be considered as conducive to the promotion of such a program." The commission agreed with Quinnell and directed him to write letters to the BPR and the state's congressional delegation expressing that fact. It was not until three years later, however, that the debate about two lane versus four lane heated up. In

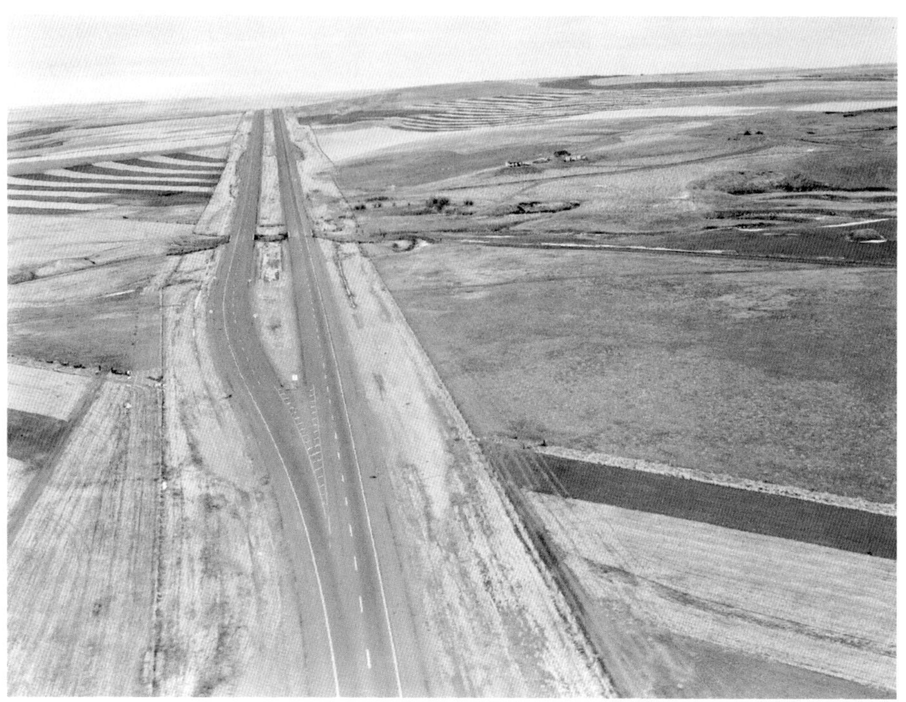

This photograph of I-15 between Cascade and Ulm shows the transition from a two-lane to four-lane controlled-access highway. *MHS Photograph Archives, Helena, PAc 86-15.154-A.*

June 1966, when Congress debated the two-lane-versus-four-lane issue, the highway commission adopted a resolution that supported the construction of *all* interstates as four-lane facilities. In September 1966, President Lyndon Johnson signed the legislation stipulating a four-lane interstate system.

With the passage of the new highway legislation, the highway commissioners and highway department planned for the expansion of the state's interstate system to four lanes. For Montana, the expansion meant that an additional $125 million would be needed to accomplish the goal through the fiscal year 1971. The state had to raise an additional $3 million per year to match federal funds so that 271 miles of two-lane highway already constructed could be added to the 638 miles of four-lane highways yet to be built.

The Interstates, 1968–1988

During the twenty-year period between 1968 and 1988, when Montana's interstate system was completed, the federal government enacted eight Federal Aid road acts that significantly impacted the pace of interstate construction, took into account the effect of the interstates on metropolitan areas and expanded the role of the public in the planning process. In addition, because of the National Environmental Protection Act (NEPA), planning and designing highways became a multidisciplinary process that took into account the impact of the interstates on environmental and social issues.

Significant changes to the interstate program began with the Federal Aid Highway Act of 1968. The act reflected President Lyndon Baines

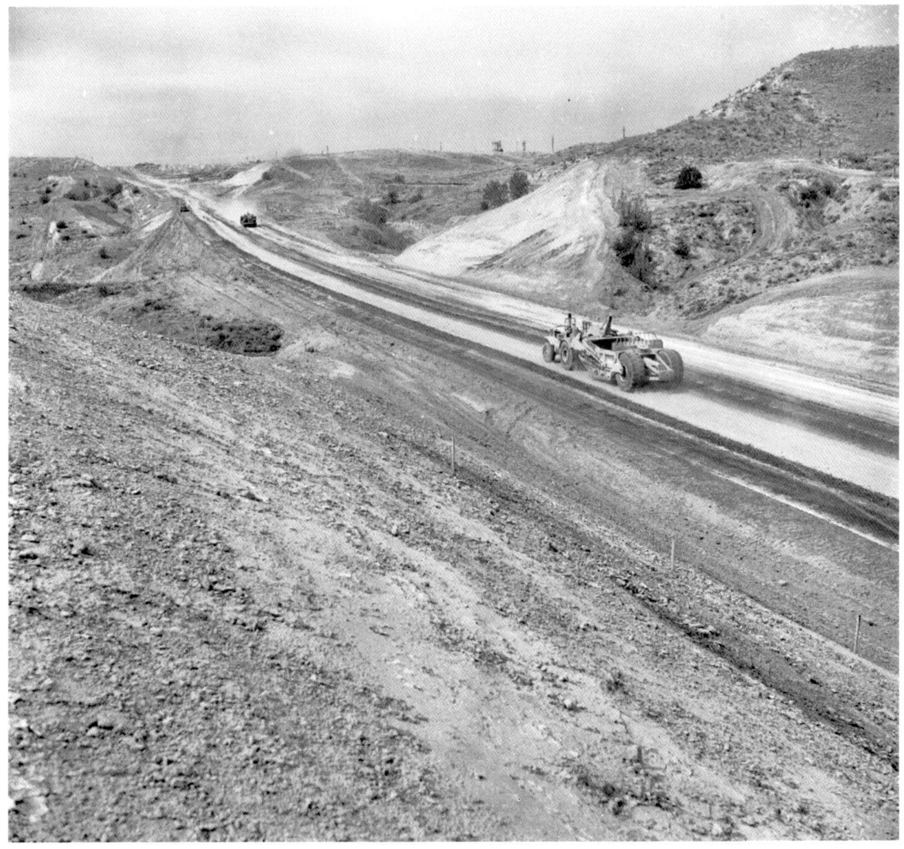

When it was completed in 1962, *Parade* magazine nominated this segment of I-94 near Miles City as "one of the nation's best new highways." *MHS Photograph Archives, Helena, PAc 86-15.370-H.*

Johnson's Great Society reforms by giving American citizens a greater voice in how Congress allocated and spent federal funds. For the most part, the federal legislation focused on the effects of the interstates on urban areas and on minority groups in the inner cities. The 1956 legislation creating the interstate program stipulated that one public hearing be held for each highway project. Public hearings were oftentimes held after the designers had determined an alignment for the superhighway. The highway departments would listen to comments made by the public and sometimes tweak designs based on those comments, but more often than not, the alignment and design were "done deals" frequently decried by those attending the hearings.

The 1968 Federal Aid Highway Act required the Federal Highway Administration (FHWA) and the state highway departments to study the social and environmental impacts of the interstates on urban environments. While Montana was primarily a rural state with most of the interstates constructed around the state's few metropolitan areas (i.e. Billings, Butte, Great Falls and Missoula) early in the interstate era, the act's requirement that the highway department hold two public hearings per interstate project affected the interstate planning process. Under the new regulations, "anyone unhappy with a state highway decision could appeal the matter directly to the federal government." Predictably, the new public hearing process was not very popular with the state highway departments as it essentially took away many design decisions from the engineers.

When President Richard Nixon signed the legislation enacting the National Environmental Policy Act (NEPA) in January 1970, few could have anticipated the long-range impact it would have on the state highway department and the interstate program. Called by some the "environmental Magna Carta," the act ensures that federal and state agencies that utilize federal funds weigh potential impacts to the environment during the decision-making process. That process is represented by a Categorical Exclusion, Environmental Assessment or Environmental Impact Statement. NEPA also instituted a multidisciplinary approach to identifying and assessing potential environmental impacts in the planning stage of federally funded projects. Consequently, the Montana Department of Highways employed biologists, archaeologists, wetland biologists and historians to identify those resources and assess any effects in consultation with other federal and state agencies. For the most part, NEPA is the umbrella for a variety of federal regulations, including the Threatened and Endangered Species Act, the National Historic Preservation Act, the Clean Air Act and others. Beginning in the early 1970s, NEPA played an expanding role in how the Montana Department

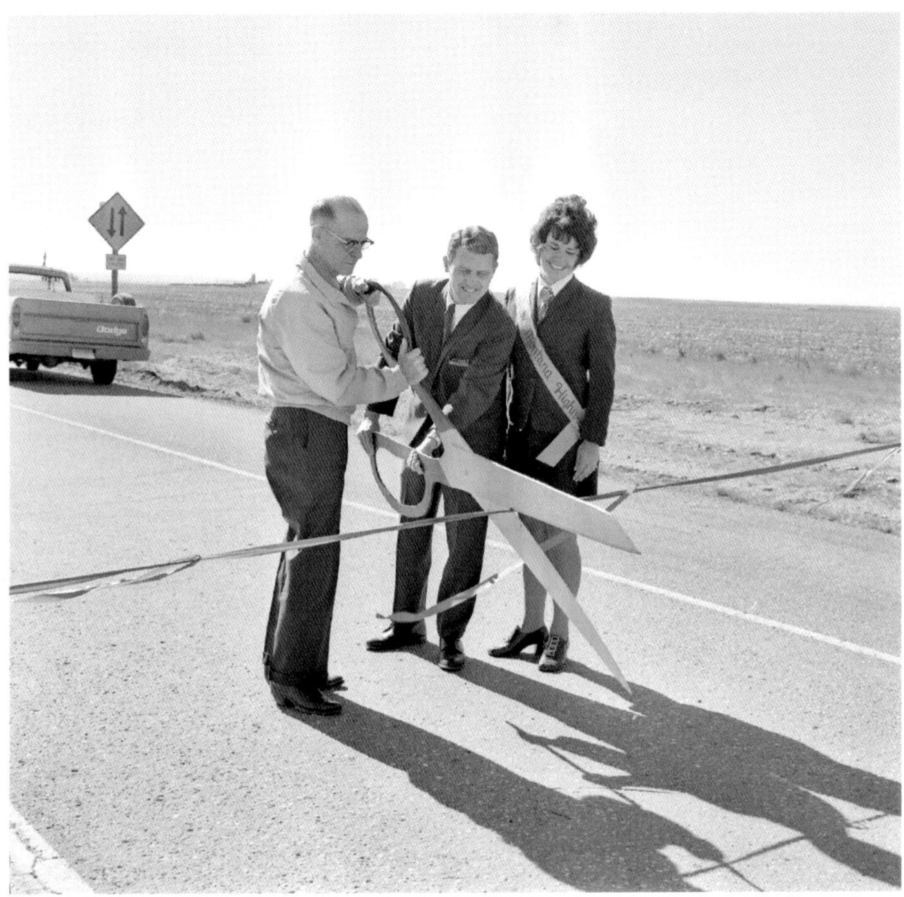

As interstate segments were completed, state and local dignitaries dedicated them with speeches and ribbon cuttings. *MHS Photograph Archives, Helena, PAc 86-15.6182.*

of Highways conducted its business. Because of NEPA, planning, designing and building highways (including the interstates) became a multidisciplinary process and was no longer purely the domain of the engineers.

Along with increased public involvement in interstate planning, design and construction came a ramped-up highway beautification and rest area construction program associated with the highways. Highway beautification was one of LBJ and his wife Lady Bird's pet projects to clean up the roadsides, remove unsightly billboards and plant trees and other vegetation along the interstates to lessen their impact on the environment. Congress passed the Highway Beautification Act in 1965 and expanded it under Richard Nixon's administration in 1970. The legislation contained three parts that involved roadside billboards, junkyard control and scenic enhancements. The act required the state legislatures to pass rules that required the removal of billboards, the screening or removal of roadside junkyards and the establishment of scenic easements adjacent to highways. In return, the federal government would fund 75 percent of the costs of the program. If states chose not to implement the beautification statutes, they could lose 10 percent of their federal highway allocation.

When Montana's interstate system was completed in 1988, the state had 1,188.7 miles of interstate highways. The longest was Interstate 90 at 542.6 miles, with I-15 and I-94 following at 395.2 miles and 247.8 miles, respectively. There were also two connector interstates: I-115 at Butte and I-315 at Great Falls. Between 1968 and 1988, the bulk of the superhighway construction was on Interstate 90. The last section of interstate completed was a little over 7.0 miles of I-15 in Beaverhead County, 9.0 miles south of Dillon. When completed, Montana's interstate highway system cost $1.22 billion, with the average cost per mile at $1.03 million. At nearly 95 percent, Montana has the highest proportion of rural interstate in the United States. The completion of the interstate highways marked the culmination of over a century of efforts by Montanans to provide an efficient and modern highway system for the state.

Since 1988, the Montana Department of Transportation (MDT) has maintained the state's highway system, reconstructed segments of functionally obsolete highways in need of improvements, replaced substandard bridges and built projects designed to improve the safety of Montana's highways. The roads have been enhanced by adjacent bicycle trails, landscaping and the preservation of wetlands, historic sites and rest areas. The highways' impact on wildlife has also been an ongoing concern to the agency, which recently included facilities to allow wildlife to cross the highway unimpeded by heavy traffic and other barriers. The history of highways is one of continual evolution as public needs change and technology improves. Nelson Parson, Michael King and Warren Gillette could hardly have envisioned the wild ride highways have taken us on since the early days of transportation in Montana.

NOTES

Introduction

1. Moulton et al., *Journals of Lewis & Clark*, 401, 407, 425.

Part I

2. Mullan, *Construction of a Military Road*, 49.
3. Ibid., 22–23; Hewitt, *Across the Plains*, 371.
4. Madsen and Madsen, *North to Montana!*, 247.
5. Burlingame, *Montana Frontier*, 137.
6. Axline, *Conveniences Sorely Needed*, 17–19; [Virginia City] *Montana Post*, February 22, 1865; January 16, 1866.
7. *Helena (MT) Tri-Weekly Republican*, July 26, 1866.
8. Axline, "Frenchwoman's Road."
9. Ibid.
10. *Omaha (NE) Herald*, October 3, 1877.
11. James Ashley in *Contributions to the Historical Society*, vol. 6, 279; Madsen and Madsen, *North to Montana*, 145–46.

Part II

12. *Senate Journal of the Thirteenth Session of the Legislative Assembly*, 165, 375, 387, 450, 464.
13. Ibid.
14. *Daily Missoulian*, January 24, 1914.
15. Ibid.
16. *First Biennial Report*, 11.
17. Wyss, *Roads to Romance*, 60–61.
18. Ibid.
19. Edy, "Summary of Work Done," 2–3.
20. *Dillon (MT) Examiner*, December 15, 1915; January 21, 1916.

Part III

21. Edy, "Road Building Record."
22. *Third Biennial Report*, 42.
23. Rader, "Montana's Highway Program," 148.
24. *Third Biennial Report*, 7.
25. *Mineral County Press*, August 16, 1917.
26. Montana Governors' Papers, Box 45, File 4.
27. Ibid.
28. Montana State Highway Commission Meeting Minutes (hereafter MSHC), Book 3, 187.
29. Montana Governors' Papers.
30. Ibid., Box 45, File 3.
31. Rader, "Montana's Highway Program," 148.
32. MSHC, Book 4, 149.

Part IV

33. *Hysham (MT) Midland Empire*, November 23, 1935.
34. Birney, *Roads to Roam*, 162.
35. Fletcher, *Headin' for the Hills*.
36. In 1939, the federal government rechristened the Bureau of Public Roads as the Public Roads Administration. Ten years later, in 1949, it changed the agency's name back to the Bureau of Public Roads, a name it would retain until 1966, when it became the Federal Highway Administration. MSHC, Book 8, 179–80.

Part V

37. MSHC, Book 9, 38.
38. FHA, *American Highways*, 275.
39. Ibid., 274.
40. [Helena, MT] *Independent Record*, April 16, 1945.
41. MSHC, Book 12, 78–79.
42. Ibid., Book 10, 10–11.
43. Planning Survey Division, *Montana Highway History*, 17.
44. MSHC, Book 12, 71; Book 13, 209–10.
45. "Montana's White Crosses," www.visitmt.com.

Part VI

46. Kologi interview.
47. Ibid.
48. *Center Line*, April 8, 1959.
49. *Independent Record*, January 25, 1963.
50. Ibid., January 29, 1963.
51. Ibid., February 12, 1963.
52. Meeting Notes, 1ff.
53. *Great Falls (MT) Tribune*, July 15, 1962.

BIBLIOGRAPHY

Books

Axline, Jon. *Conveniences Sorely Needed: Montana's Historic Highway Bridges, 1860–1956.* Helena: Montana Historical Society Press, 2005.

Bateman, John H. *Introduction to Highway Engineering: A Textbook for Students of Civil Engineering.* 3rd ed. New York: John Wiley & Sons, Inc., 1939.

Birney, Hoffman. *Roads to Roam.* Philadelphia: Penn Publishing Co., 1930.

Burlingame, Merrill G. *The Montana Frontier.* Helena, MT: State Publishing Company, 1942.

Contributions to the Historical Society of Montana. Vol. 6. Helena: Historical Society of Montana, 1907.

———. Vol. 8. Helena: Montana Historical and Miscellaneous Library, 1917.

Federal Highway Administration. *America's Highways, 1776–1976: A History of the Federal-Aid Program.* Washington, D.C.: Government Printing Office, 1976.

Fletcher, Robert H. *Headin' for the Hills.* Helena: Montana State Highway Department, 1937.

Hewitt, Randall H. *Across the Plains and Over the Divide.* New York: Broadway Publishing, 1906.

Jackson, W. Turrentine. *Wagon Roads West: A Study of Federal Road Surveys and Construction in the Trans-Mississippi West, 1846–1869.* Lincoln: University of Nebraska Press, 1964.

Lewis, Tom. *Divided Highways: Building the Interstate Highways, Transforming American Life.* New York: Viking, 1997.

Madsen, Betty M., and Brigham D. Madsen. *North to Montana! Jehus, Bullwhackers, and Mule Skinners on the Montana Trail.* Logan: Utah State University Press, 1998.

Malone, Michael P., Richard B. Roeder and William L. Lang. *Montana: A History of Two Centuries.* Rev. ed. Seattle: University of Washington, 1991.

Moulton, Gary E., et al. *The Definitive Journals of Lewis & Clark.* Vol. 3, *Up the Missouri to Fort Mandan.* Lincoln: University of Nebraska Press, 1987.

Planning Survey Division. *Montana Highway History*. Vol. 2, *1943 to 1959*. Helena: Montana State Highway Commission, 1960.

Quivik, Fredric L. *Historic Bridges in Montana.* Washington, D.C.: Department of the Interior, 1982.

Schwantes, Carlos Arnaldo. *Long Day's Journey: The Steamboat & Stagecoach Era in the Northern West.* Seattle: University of Washington, 1999.

Senate Journal of the Thirteenth Session of the Legislative Assembly of the State of Montana. Helena, MT: Independent Publishing Co., 1913.

Statewide Highway Planning Survey. *History of the Montana State Highway Department, 1913–1942.* Helena: Montana State Highway Commission, 1943.

Swift, Earl. *The Big Roads: The Untold Story of the Engineers, Visionaries, and Trailblazers Who Created the American Superhighways.* New York: Houghton Mifflin Harcourt Publishing, 2011.

Wyss, Marilyn. *Roads to Romance: The Origins and Development of the Road and Trail System in Montana.* Helena: Montana Department of Transportation, 1992.

Magazines and Journals

Axline, Jon. "Building Permanent and Substantial Roads: Prison Labor on Montana's Highways, 1910–1925." *Montana: The Magazine of Western History* 63, no. 2 (Summer 2012).

———. "A Massive Undertaking: Constructing Montana's Interstate Highways, 1956–1988. *Montana: The Magazine of Western History* 63, no. 3 (Autumn 2013).

Fisher, W.D. "Good Roads in the West." *Good Roads* 66 (January 1924).

Rader, R.D. "Montana's Highway Program and Problems." *Western Construction News* 3 (March 10, 1928).

Personal Interviews

Robert Champion, July 2013.

Lewis Chittim, April 1996.

Robert Dunbar, April 1996.

Steve Kologi, April 1996; July 2013.

Other

Axline, Jon. "The Frenchwoman's Road." In "More from the Quarries of Last Chance Gulch." *Helena (MT) Independent Record*, June 10, 1999.

Center Line, 1938–40, 1958–65. Montana Historical Society Research Center. Helena, MT.

Edy, John. "Road Building Record Made During 1920." *Fergus County Argus*, September 17, 1920.

Meeting Notes, Interstate Project No. I 15-4(3)209. Wolf Creek to Sieben, January 19, 1961; Transcript of a Public Hearing Involving a Highway Construction Project on Interstate Route #15 Between Sieben and Wolf Creek (Project I 15-4(3)). Prepared by the Planning Survey Division, Montana Highway Commission, March 24, 1961.

Montana Governors' Papers. Manuscript Collection 35. Research Center. Montana Historical Society. Helena, MT.

Montana State Highway Commission Meeting Minutes, 1913–1988.

Wohlgenant, Carl F., Jr. "Development of the Federal Aid Highway System in Montana." Master's thesis. Montana State University, Missoula, MT.

Reports

Edy, John. "Summary of Work Done by the State Highway Commission of Montana, Montana." December 15, 1920.

Metlen, George R. "First Biennial Report Montana State Highway Commission, December 1, 1914." Montana Governors' Papers: Agency and Subject Files. Manuscript Collection 35, Box 222, Folder 17. Montana Historical Society Research Center, Helena, MT.

———. *Report of the Montana Highway Commission for the Years, 1915–1916.* Helena: Montana State Highway Commission, 1917.

Mullan, John. *Report on the Construction of a Military Road from Fort Walla-Walla to Fort Benton.* Washington, D.C.: Government Printing Office, 1863.

Report of the State Highway Commission of Montana for Biennium Ending December 1930. Helena: Montana State Highway Commission, 1930.

Report of the State Highway Commission of Montana for Biennium Ending December 1932. Helena: Montana State Highway Commission, 1932.

Report of the State Highway Commission of Montana for Period Ending December 1928. Helena: Montana State Highway Commission, 1928.

Second Biennial Report of the State Highway Commission of Montana, 1919–1920. Helena: Montana State Highway Commission, 1920.

Third Biennial Report of the State Highway Commission of Montana, 1921–1922. Helena: Montana State Highway Commission, 1922.

Newspapers

Billings Gazette
Bozeman Courier
Butte Miner
Fergus County Argus
Great Falls Tribune
Havre Daily News-Promoter, October 13, 1927.
Helena Independent
Helena Independent Record
Hysham Midland Empire
Missoulian
Montana Post
Montana Standard
Philipsburg Mail

INDEX

A

B

C

D

E

F

T

V

W

Y

ABOUT THE AUTHOR

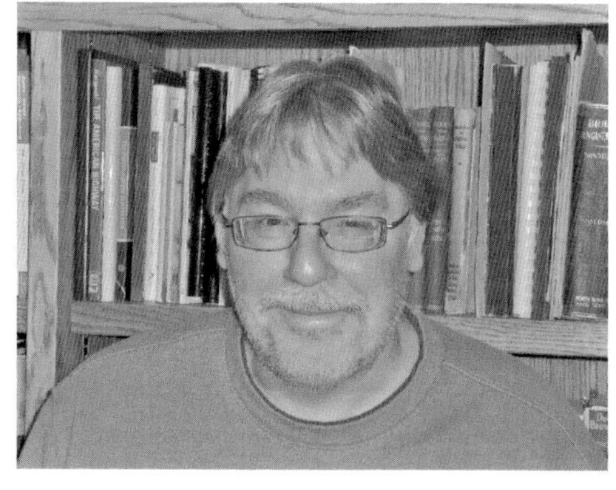

Jon Axline has been the historian at the Montana Department of Transportation (MDT) since 1990. When not sweating over the state's historic roads and bridges, he conducts cultural resource surveys and writes the MDT's roadside historical and geological interpretive markers. He is a regular contributor to *Montana: The Magazine of Western History* and *Montana Magazine*. He is also author of *Conveniences Sorely Needed: Montana's Historic Highway Bridges* and editor of *Montana's Historical Highway Markers.* Jon lives in Helena with his wife and three and a half Welsh Pembroke corgis.